NON-GRAVITATIONAL PERTURBATIONS AND SATELLITE GEODESY

NON-GRAVITATIONAL PERTURBATIONS AND SATELLITE GEODESY

ANDREA MILANI
ANNA MARIA NOBILI
PAOLO FARINELLA
*Dipartimento di Matematica,
Università di Pisa*

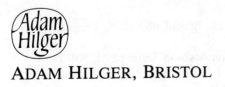

ADAM HILGER, BRISTOL

© IOP Publishing Ltd 1987

All rights reserved. No part of this publication may be reproduced, stored in a retrieval system or transmitted in any form or by any means, electronic, mechanical, photocopying, recording or otherwise, without the prior permission of the publisher.

British Library Cataloguing in Publication Data

Milani, Andrea
   Non-gravitational perturbations and
   satellite geodesy.
   1. Artificial satellites——Orbits
   I. Title   II. Nobili, Anna Maria
   III. Farinella, Paolo
   692.43'4   TL1080

ISBN 0-85274-538-9

Consultant Editors: **Professor A J Meadows**, Loughborough University
                           Professor A E Roy, University of Glasgow

Published under the Adam Hilger imprint by
IOP Publishing Ltd
Techno House, Redcliffe Way, Bristol BS1 6NX, England

Typeset by Mathematical Composition Setters Ltd, Salisbury
Printed in Great Britain by J W Arrowsmith Ltd, Bristol

Contents

1 Introduction	**1**
2 The perturbations	**6**
2.1 Satellite orbits as small force experiments	6
2.2 Orders of magnitude of the perturbing forces	14
2.3 Tides and apparent forces	20
3 Tools from celestial mechanics	**27**
3.1 Keplerian orbits	27
3.2 Variations of the elements	32
3.3 Non-singular elements	38
3.4 Secular perturbations	42
4 Solar radiation pressure: direct effects	**48**
4.1 Satellite–solar radiation interaction	48
4.2 Long-term effects on semi-major axis	51
4.3 Long-term effects on the other orbital elements	60
4.4 Conclusions and examples	67
5 Radiation pressure: indirect effects	**77**
5.1 Earth-reflected radiation pressure	77
5.2 Anisotropic thermal emission	87
5.3 Radiowave beams	93
5.4 Eclipses	94
6 Drag	**98**
6.1 Orbital perturbations by a drag-like force	98
6.2 Drag coefficients	103
6.3 Charged particle drag	109
7 Manoeuvres	**113**
7.1 Orbital manoeuvres	113
7.2 Attitude manoeuvres and thruster leakages	117
References	**120**
Index	**123**

1

INTRODUCTION

The definition of celestial mechanics as a discipline is tainted with ambiguity. Celestial mechanics traditionally refers to the gravitational N-body problem, or to its variants, that is to problems which can be studied within the formalism of Hamiltonian mechanics. On the other hand, the purpose of celestial mechanics is to explain and predict the motion of every real object in outer space, no matter what the causes of the changes in its orbit. As long as the bodies known to orbit in outer space were only the large natural celestial bodies—such as the planets—this ambiguity was of no consequence: their motion was described with astonishing accuracy by the laws of 'clean', conservative celestial mechanics.

Some exceptional cases were already known in the nineteenth century. The non-gravitational thrust experienced by some comets was discovered by Encke for the comet now bearing his name (Whipple and Sekanina 1979). The existence of a deceleration of the Moon along its orbit that is not explained by the gravitational influence of the Sun and the planets was proved by J C Adams (Smart 1947); later G H Darwin explained this effect as due to the dissipation of the energy of the tides—raised by the Moon itself—in the Earth's seas (Darwin 1908). However, these exceptions were few and the non-gravitational effects were only very small corrections to the purely gravitational theory, a theory which was as a whole exceedingly successful in describing the orbits of the planets and the natural satellites.

On 4 October 1957 all this changed. The bodies orbiting in outer space, from then on, included artificial ones. Such bodies would experience non-gravitational perturbations to their orbit of such a size that all the tricks of 'dirty' celestial mechanics were immediately needed. Of course, most of the same physical phenomena also act on the natural celestial bodies: however, one parameter controlling the size of the non-gravitational forces relative to the gravitational ones is the area-to-mass ratio, and this quantity is bound to be quite high for spacecraft (usually between 0.1 and 0.01 $cm^2 g^{-1}$), and many orders of magnitude smaller even for a tiny asteroid. As a result, non-gravitational perturbations came up as a challenge for celestial mechanicians of the time, and since the subject was new and almost unexplored 'to be young was very heaven', as Dr King-Hele puts it (by quoting Wordsworth) in his passionate account of the dawn of space age (King-Hele 1983).

The non-gravitational perturbations arise essentially from two causes. Firstly, outer space is not empty. The Earth's atmosphere extends up to much higher levels than was assumed before the space age, and because of the high relative velocity of the spacecraft with respect to the gas particles a significant aerodynamic drag arises even with a very low density. This was the first kind of non-gravitational perturbation to be investigated by analysis of the tracking data of Sputnik 1, the first paper being printed five weeks after the launch (*Nature*, 9 November 1957). Drag is the most important non-gravitational perturbation for low satellites; however, it is not the most important topic of this book, for two reasons. The amount of drag essentially depends on atmospheric density at the satellite height, and the extremely intricate subject of atmospheric models is outside the scope of this book. On the other hand, whenever atmospheric studies start, satellite geodesy more or less ends, because atmospheric drag is so difficult to model that the accuracy of the orbit determination will necessarily degrade, and unless resonance phenomena are exploited, the geodetic parameters of interest (e.g. gravitational anomalies, station coordinates, Earth orientation angles, tide coefficients) will not be recovered with the required accuracy. Therefore, unless some kind of 'drag-free' technology can be used, geodetic satellites almost always fly in quite high orbits.

Just because the accuracy of the orbit determination required by today's satellite geodesy is so high (the orders of magnitude are discussed in Chapter 2) the problem of modelling drag has been displaced to higher orbits. Before its launch (in 1976), the geodynamics satellite LAGEOS was predicted to experience a

INTRODUCTION

negligible drag, because of a very high orbit (almost 6000 km high) and a very small area-to-mass ratio. But thanks to very accurate laser tracking, and after accounting for all known perturbations, it was found that LAGEOS was experiencing a drag-like deceleration of about 3×10^{-10} cm s^{-2}. To reconcile this unexpected finding with current atmospheric models, a lot of work has been done to understand the details of the interaction of the spacecraft with a very thin atmosphere and also with the charged particles of the plasmasphere. However, this drag-like force is also variable with time in a not very predictable way, and for the specialists in celestial mechanics who were too young to exploit the 'heavenly' opportunity of Sputnik 1 (as a matter of fact, we were children), this has been the hardest challenge of modern satellite geodesy.

Outer space is also not empty because it is pervaded by electromagnetic radiation: the light arriving directly from the Sun, and that reflected by the Earth. The photons of electromagnetic radiation exchange momentum with spacecraft whenever they are absorbed or reflected; spacecraft themselves give out infrared radiation that can carry away some momentum. The resulting accelerations are small—typically of the order of 10^{-6} cm s^{-2}—but are of course not negligible whenever we want to compute an orbit with an accuracy of a few centimetres. For high orbiting satellites with complex shapes this is the main, and also the most troublesome, non-gravitational perturbation.

Two approaches to the modelling of orbital perturbations due to radiation pressure are possible. We can try to compute the effect of the interaction of every piece of the spacecraft's surface with the sunlight, and produce a very complicated computer program to be used in the orbital computations. However, this purely numerical procedure tends to give an illusory precision, and when the uncertainties in the properties of the satellite surface and its orientation in space are properly taken into account, it is found that the real model accuracy never exceeds a few per cent.

The other approach is to use the classical tools of celestial mechanics to find which component of the radiation pressure acceleration is really important, because its effects accumulate with time. It is found that the main perturbative effect can be produced by a minor fraction of the total perturbing force; a typical result says that the knowledge of a small misalignment angle of an antenna can be the only relevant information in order to model in an effective way the long-term perturbations. An even more complicated problem is posed by earthshine, for the very simple reason that the Earth's surface elements reflect sunlight in a way changing with time and with the geographical features. This problem is again too complicated to be tackled in a purely numerical way, unless it is first simplified by some analytical theory that shows which are the really relevant features.

Another reason why non-gravitational perturbations are unavoidable in satellite geodesy is because artificial satellites are not launched for the sake of celestial mechanicians who like to compute their orbits; to this rule there are a few, although very important, exceptions. This implies that almost all satellites have a power system that transforms absorbed sunlight and uses the energy collected at the surface of the spacecraft through its body, eventually re-emitting it either as thermal radiation or as radiowaves beamed towards the Earth. This makes the problem of the exchange of momentum with radiation much more complicated. Moreover, to increase the capability of modern spacecraft, the power system is augmented by adding steerable wings covered by photovoltaic cells and the antennae are made steerable and with complex shapes to be highly directional; when it comes to an accurate orbit computation, all these appendages are a real nightmare both because of radiation pressure and because of drag, at least for low satellites.

Almost all spacecraft now have an attitude and orbit control system; this in turn will perturb the orbit, sometimes on purpose (orbital manoeuvres), sometimes as a by-product (attitude manoeuvres), sometimes by leaks from the fuel tanks. One might ask why we do not restrict our analysis to the simpler cases of passive or at least simple satellites. The reason lies in what experience of satellite geodesy has taught us, namely that it is not wise to wait for an *ad hoc* satellite to look for relevant geophysical information. This was true for the very earliest satellites that preceded by years the dedicated geodetic satellites, and is true today with the soaring costs of sophisticated single-purpose space missions; at the very best, a space mission will be designed as a compromise between different purposes, the geodetic use being only one of these.

The purpose of this book should then be clear: we want to enable readers to assess for themselves the possible geodetic uses of a given space mission, or maybe to propose a new one (or more likely to propose a combined geodetic use for a mission envisaged for other purposes as well). This is indeed what we have been doing in the last six years. Although other sections of the knowledge necessary for this purpose have to be found in other textbooks, such as Kaula's (1966) and Lambeck's (1980), we hope to be able to give the basic ideas on the physics of the main non-gravitational perturbations, and the mathematics of the methods to compute their orbital effects. Our purpose is not that of giving an exhaustive treatment of each particular problem, but rather to use typical examples to convey to the reader a feeling of the relevance of the different problems that are to be solved in order to achieve a given level of accuracy in the orbit determination and in the recovery of geophysically significant parameters.

ACKNOWLEDGMENTS

We have discussed this subject, and in this way received significant input, with so many people that it would be difficult to thank them all. However, we would like to mention our co-workers L Anselmo, M Carpino and F Sacerdote in Pisa, and B Bertotti, of the University of Pavia. We also thank A T Sinclair of the Royal Greenwich Observatory, UK, F Barlier and F Mignard of CERGA, France, and C A Wagner of NOAA, USA. Special thanks are due to our friend Archie E Roy of the University of Glasgow, who kindly offered to edit the book and improve the English. Finally, this book is an occasion to remember Professor Giuseppe Colombo, recently passed away, who first taught us that there is neither 'clean' nor 'dirty' celestial mechanics, but only real problems to be solved in the sky.

2

THE PERTURBATIONS

2.1 SATELLITE ORBITS AS SMALL FORCE EXPERIMENTS

A satellite orbiting the Earth feels accelerations due to a wide range of physical causes, resulting from the interactions of the spacecraft with natural celestial bodies and with the atoms, particles and fields encountered along its way. Of course some of these accelerations are much smaller than others, and some can be entirely neglected, but to decide which are negligible is not straightforward; forces are negligible not in themselves, but with respect to other forces or with respect to the achievable accuracy in the orbit determination. The latter is also a function of the accuracy of the observations of the spacecraft position. It is precisely because of the remarkable increase in accuracy of distance measurements from the satellite to points on the ground that the need has arisen for a very refined study of the orbital perturbations.

To give an idea of the accuracies that are involved we consider three typical examples of today's (1986) technology in satellite geodesy. Satellite laser ranging (SLR) directly observes the distance from a ground station to the spacecraft by measuring the two-way travel time of a laser pulse. In the most advanced stations the pulses are 6 cm long, and the system measuring the starting and the return time is accurate at this level (i.e. about 10^{-10} s). The results look more or less as drawn in figure 2.1, where only the residuals, i.e. the differences between the predicted and the observed distances to the satellite, are plotted. Figure 2.1 shows that when the orbit is predicted (usually many months in advance for LAGEOS) this prediction fails by an amount that is not large in itself (a few tens of metres in this case) but is nevertheless much larger than the scatter of the observation points around a well defined curve; since we can assume that the scatter is due to the accidental errors of the observing apparatus, this amounts to saying that the accuracy of the model of the satellite orbit fails to cope with the accuracy of the observations. As will be discussed in Chapters 5 and 6, in the special case of the dedicated geodetic satellite LAGEOS the physical cause of this prediction error is a combination of non-gravitational forces that are very difficult to predict in advance.

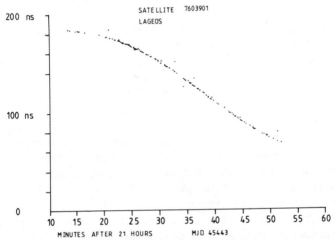

Figure 2.1 A pass of the geodetic satellite LAGEOS over the Royal Greenwich Observatory SLR station at Herstmonceux, East Sussex, UK. The vertical scale is in nanoseconds of two-way travel time, each corresponding to 15 cm in the station–satellite distance. Once a few obviously spurious points have been eliminated, the curve can be fitted to a suitable polynomial; this fitting leaves residuals with a root mean square of about 5 cm. (Courtesy of A T Sinclair.)

The Global Positioning System (GPS) is a network of satellites and ground stations of the US navy. Every satellite broadcasts a coded radiowave beam, and by receiving and comparing the signals from many satellites the position of an aircraft or a ship can be determined with an accuracy of a few metres. Apart from obvious military or navigational uses, the GPS can be used to measure the vector from a ground station to a spacecraft (or something equivalent) with greater accuracy, down to a few centimetres. However, this does not allow geodetic positioning in a global reference frame with the same accuracy: satellites wander around their computed orbits, because of non-modellable, non-gravitational perturbations, by a few metres. This example is further discussed in Chapters 4 and 5.

Radar altimeter satellites, such as the American spacecraft SEASAT flown in 1978 and the European project ERS-1, can measure the height of the satellite above the mean sea surface with an accuracy of about ten centimetres. Updated radar technology could give even better accuracy, but it would be very difficult to exploit this because these satellites use very powerful radiowave beams, therefore they must have large orientable solar panels and large radar antennae (see figure 2.2). These large surfaces with their variable attitude produce relatively large non-gravitational perturbations that are very difficult to model, and again the orbit cannot be determined with one-centimetre accuracy even if the best tracking methods are used. This example is further discussed in Chapters 4 and 6. Since accuracies of one centimetre (or less) in the sea surface height would represent a dramatic improvement both in gravimetry (through measurement of the geoid) and in oceanography (by measuring the sea-surface topography, i.e. the differences between the mean sea surface and the geoid), two solutions are being studied. The first is based on 'drag-free' technology: whenever the spacecraft feels an external acceleration that is non-gravitational, an on-board accelerometer can measure it and the orbit can be corrected with the action of thrusters to compensate these sensed accelerations (alternatively, the accelerations only can be measured and the compensation can be done in the orbit computation process). The second solution is based on the idea of tracking the altimeter satellite in an essentially continuous way, thus relying very little on the computation of the orbit based on a force model. Both methods in effect abandon the idea of modelling non-gravitational perturbations; however, the technologies involved are so expensive that it is better ... to read this book.

Moreover, if anybody reads this book in ten years' time, this discussion on satellite geodesy technologies will probably look very out of date. Time and frequency measurement technology, and, as its corollary, distance and relative velocity measurement technology, have been progressing at such a pace in the last 30 years

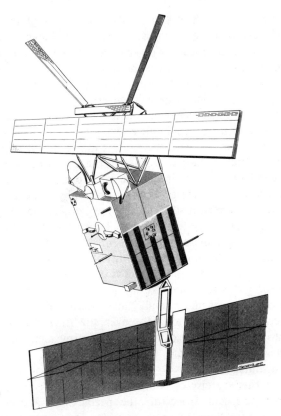

Figure 2.2 The ERS-1 altimetry satellite, a European Space Agency project to be launched in 1989. Note the complexity of the shape resulting from steerable solar panels and microwave antennae. (Courtesy ESA.)

that it is unlikely they will suddenly come to a standstill in the near future. We have to prepare ourselves with all the mathematical, physical and computer science tools available, for the subcentimetre accuracy distance measurements of the future. On the contrary, the state of the art in satellite geodesy is such that non-gravitational force models often represent the main limitation to accurate orbit determination, as in the examples above. (We do not want to understate the complexity of the gravimetry–altimetry–station positioning problem even without non-gravitational perturbations; studies on tides and reference systems are also at the cutting edge of research. We wish only to stress that all the relevant problems have to be solved together to some degree to take full advantage from the accuracy of space geodesy measurements.)

The purpose of this chapter is therefore to display a list of possible physical mechanisms affecting satellite orbits, and to compute for each one an order-of-magnitude estimate of the perturbing accelerations, because this is of course the first step to be taken to assess what is or is not negligible and what is or is not a problem. The reader unfamiliar with the subject might be surpised by the exceedingly small values of some of these estimates: however, satellite geodesy is the discipline that tries to measure the smallest accelerations ever considered as objects of a physical experiment— with LAGEOS we are currently (late 1986) working at the 10^{-10} cm s^{-2} level, at least for long-term effects. Each satellite orbit can be thought of as a small force experiment, where the (almost) void outer space provides the most favourable laboratory to measure accelerations that would go undetected in the noisy environment of an Earth-bound laboratory. To decide that a physical interaction is negligible we must find a small value indeed for the acceleration it produces on the spacecraft. Moreover, the list of the forces to be taken into account increases with every improvement in measurement accuracy and coverage.

The formulae given in §2.2 should enable readers to do for themselves this exercise of compiling a list of effects to be considered for the specific spacecraft, the specific orbits and the specific accuracy requirements of space missions they are interested in. However, we have also included tables with numerical values for four specific examples we have chosen among the most representative cases of today's satellite geodesy. Example 1 is a geosynchronous spacecraft with a structure more or less typical for a small telecommunications or navigation satellite. The main advantage of geosynchronous satellites for geodesy is that there are so many of them available, and just to track them from the ground is a very inexpensive way to perform a small force experiment which can reveal significant geophysical phenomena, such as secular changes of the geopotential with time (Catalano *et al* 1983). Example 2 is LAGEOS (see figure 2.3); it is a completely passive satellite, covered with 'corner cube' mirrors to reflect back the laser pulses, and its only purpose is to allow us to compute its orbit. Example 3 is another passive laser-reflector satellite, the French 'Starlette';

SATELLITE ORBITS AS SMALL FORCE EXPERIMENTS 11

smaller and lower flying than LAGEOS, Starlette feels more the effects of atmospheric drag, tides and gravitational anomalies. Example 4 is a representative example of the altimeter satellite class, such as SEASAT or the future ERS-1.

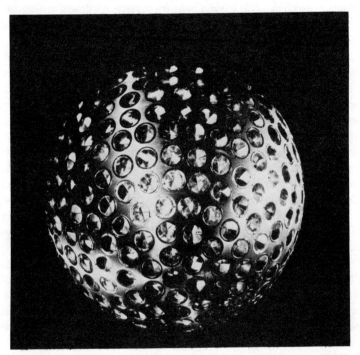

Figure 2.3 LAGEOS geodynamics satellite, with 426 retroreflectors. LAGEOS measures 60 cm in diameter and weighs 411 kg. It is at an altitude of about 6000 km and orbits the Earth in 3 h 25 min. (Courtesy NASA.)

Readers will notice that these tables, as well as the discussion in §2.2, contain non-gravitational as well as gravitational perturbing accelerations; the latter are there for comparison purpose, and are scarcely mentioned in the other chapters of this book. We apologise for a very simplified treatment of gravitational anomalies, relativistic effects and other interesting topics that are outside the scope of this book.

2.2 ORDERS OF MAGNITUDE OF THE PERTURBING FORCES

In table 2.1 we list all the relevant sources of perturbation which affect the motion of artificial satellites employed for the purposes of space geodesy. The second column gives an order-of-magnitude formula for the acceleration due to a given physical cause, and the third column lists the parameters which must be known in advance (or determined by differential corrections) to obtain an accurate dynamical model, together with their current or estimated values. These values are then inserted in the formulae to calculate the accelerations 'felt' by the four satellites quoted earlier (for them the semi-major axis a and the area-to-mass ratio $\mathcal{A}/\mathcal{M}$ are given in the first row of the table). In table 2.2 we review the parameters whose present uncertainty is capable of degrading the accuracy of the dynamical models. The third column of the table shows the (relative) estimated uncertainties of the different parameters, and the following columns display the corresponding accelerations for the four 'sample satellites', giving an idea of the achievable levels of accuracy and of the most critical problems arising when an accurate orbital propagation is needed.

In the following we present some brief comments on the various phenomena taken into account in the tables (note that the labels on the tables' rows correspond to the numbers of the following list of comments), explaining at the same time the motivation of our estimates and giving appropriate references for the most complex problems.

(1) The Earth's monopole term, $GM_\oplus/r^2$ (G is the gravitational constant, $M_\oplus$ is the mass of the Earth, and r is the distance between the Earth's centre and the satellite), is obviously the largest force which influences the satellite's motion. The corresponding acceleration can be taken as the fundamental quantity to which the various perturbing effects must be compared in order to evaluate how much they 'shift' the satellite's motion with respect to a Keplerian orbit. Thus, for instance, a perturbing acceleration of 10^{-6} cm s^{-2} on LAGEOS will cause a fractional change in the satellite semi-major axis of about $10^{-6}/3 \times 10^2 \simeq 3 \times 10^{-9}$ (i.e. an absolute change of about 4 cm) on a time-scale comparable with the orbital period (the long-term behaviour of the perturbation is determined by other factors, as we shall see in Chapter 3). As for

ORDERS OF MAGNITUDE OF THE PERTURBING FORCES

the uncertainty of $GM_\oplus$, it is easy to realise (for instance by recalling Kepler's third law) that the determination of $GM_\oplus$ requires simultaneous measurement of the orbital period (which is straightforward) and of the semi-major axis. With current tracking methods this latter always involves in some way knowledge of the velocity of light, for which a conventional value is usually chosen. Thus, the uncertainty stated in table 2.2, which corresponds to the difference between the GEM-L2 (Lerch *et al* 1983) and the GEM 9 (Lerch *et al* 1979) geopotential models, is directly connected with the uncertainty in range of the tracking methods: 10 cm over 10 000 km means an uncertainty of the order of 10^{-8} in semi-major axis, and an uncertainty three times larger (by Kepler's third law) in $GM_\oplus$. However, the uncertainties shown in the table do not have the bad consequences that one might imagine from their magnitude, because with current tracking data (range, range rate, etc) in the differential correction procedure an error in $GM_\oplus$ can be compensated almost perfectly by an error in the semi-major axis of the orbit.

(2) The distribution of mass within the Earth is not spherically symmetric, and departures from symmetry cause additional forces to appear with respect to the Newtonian (or monopole) term. Usually the gravitational potential V of the Earth (the geopotential) is expanded into a series of spherical harmonics of the form

$$V(r, \lambda, \varphi) = \frac{GM_\oplus}{r}\left[1 + \sum_{l=2}^{\infty}\sum_{m=0}^{l}\left(\frac{R_\oplus}{r}\right)^l P_{lm}(\sin\varphi) J_{lm} \cos m(\lambda - \lambda_{lm})\right] \quad (2.1)$$

where $R_\oplus$ is the mean equatorial radius of the Earth, φ and λ are the satellite's geocentric latitude and longitude, $P_{lm}(\sin\varphi)$ are the associated Legendre functions and the coefficients J_{lm} and λ_{lm} depend on the Earth's mass distribution (see Kaula 1966). The indices l and m are called, respectively, degree and order of a particular harmonic term; note that for each order l one has in general $(l+1)$ perturbing terms, so that to evaluate the acceleration caused by *all* the terms of a given order we should multiply the effect of a single harmonic by $(l+1)^{1/2}$ (assuming that the various terms are uncorrelated and add together in a random-walk way). Since the magnitude of the Legendre functions (and hence of the coefficients) varies greatly with l and m, in order to make the coefficients more

Table 2.1 Accelerations on spacecraft used for satellite geodesy.

Cause	Formula	Parameters (in CGS units)	Geosynchronous satellite $a = 42\,160$ km $\mathcal{A}/\mathcal{M} = 0.1$ cm^2 g^{-1}	LAGEOS 12 270 0.007	Starlette ~7300 0.01	SEASAT (or ERS-1) ~7100 0.2
			Accelerations (cm s^{-2})			
(1) Earth's monopole	$\dfrac{GM_\oplus}{r^2}$	$GM_\oplus = 3.986 \times 10^{20}$	2.2×10^1	2.8×10^2	7.5×10^2	7.9×10^2
(2) Earth's oblateness	$3\dfrac{GM_\oplus}{r^2}\left(\dfrac{R_\oplus}{r}\right)^2 \bar{J}_{20}$	$\bar{J}_{20} = 4.84 \times 10^{-4}$ $R_\oplus = 6.378 \times 10^8$	7.4×10^{-4}	1.0×10^{-1}	8.3×10^{-1}	9.3×10^{-1}
(2) Low-order geopotential harmonics: e.g. $l = 2$, $m = 2$	$3\dfrac{GM_\oplus R_\oplus^2}{r^4}\bar{J}_{22}$	$\bar{J}_{22} = 2.81 \times 10^{-6}$	4.3×10^{-6}	6.0×10^{-4}	4.8×10^{-3}	5.4×10^{-3}
$l = 6$, $m = 6$	$7\dfrac{GM_\oplus R_\oplus^6}{r^8}\bar{J}_{66}$	$\bar{J}_{66} = 2.42 \times 10^{-7}$	4.5×10^{-10}	8.8×10^{-6}	5.6×10^{-4}	7.0×10^{-4}
(2) High-order geopotential harmonics: e.g. $l = 18$, $m = 18$	$19\dfrac{GM_\oplus R_\oplus^{18}}{r^{20}}\bar{J}_{1818}$	$\bar{J}_{1818} = 1.8 \times 10^{-8}$	1.3×10^{-20}	6.9×10^{-10}	2.2×10^{-5}	3.9×10^{-5}
(3) Perturbation due to the Moon	$2\dfrac{GM_{\mathrm{C}}}{r_{\mathrm{C}}^3}r$	$M_{\mathrm{C}} = M_\oplus/81.3$ $r_{\mathrm{C}} = 3.8 \times 10^{10}$	7.3×10^{-4}	2.1×10^{-4}	1.3×10^{-4}	1.3×10^{-4}

ORDERS OF MAGNITUDE OF THE PERTURBING FORCES

(3) Perturbation due to the Sun	$2\dfrac{GM_\odot}{r_\odot^3}r$	$M_\odot = 3.29 \times 10^5 M_\oplus$ $r_\odot \simeq 1.5 \times 10^{13}$	3.3×10^{-4}	9.6×10^{-5}	5.7×10^{-5}	5.6×10^{-5}
(3) Perturbation due to other planets (e.g. Venus)	$2\dfrac{GM_\varphi}{r_\varphi^3}r$	$M_\varphi = 0.82 M_\oplus$ $r_\varphi \gtrsim 4 \times 10^{12}$	4.3×10^{-8}	1.3×10^{-8}	7.5×10^{-9}	7.3×10^{-9}
(4) Indirect oblation	$3\bar{J}_{20}\dfrac{GM_\oplus}{r_\mathbb{C}^2}\left(\dfrac{R_\oplus}{r_\mathbb{C}}\right)^2\dfrac{M_\mathbb{C}}{M_\oplus}$		1.4×10^{-9}	1.4×10^{-9}	1.4×10^{-9}	1.4×10^{-9}
(5) General relativistic correction	$\dfrac{GM_\oplus}{r^2}\dfrac{GM_\oplus}{c^2}\dfrac{1}{r}$	$\dfrac{GM_\oplus}{c^2} = 0.44$	2.3×10^{-9}	9.5×10^{-8}	4.5×10^{-7}	4.9×10^{-7}
(6) Atmospheric drag	$\dfrac{1}{2}C_D\dfrac{\mathscr{A}}{\mathscr{M}}\rho V^2$	$C_D = 2\text{--}4$ $\rho = 0\text{--}10^{-16}$	$0(?)$	3×10^{-10}	7×10^{-8}	2×10^{-5}
(7) Solar radiation pressure	$\dfrac{\mathscr{A}}{\mathscr{M}}\dfrac{\Phi_\odot}{c}$	$\Phi_\odot = 1.38 \times 10^6$	4.6×10^{-6}	3.2×10^{-7}	4.6×10^{-7}	9.2×10^{-6}
(8) Earth's albedo radiation pressure	$\dfrac{\mathscr{A}}{\mathscr{M}}\dfrac{\Phi_\odot}{c}A(\oplus)\left(\dfrac{R_\oplus}{r}\right)^2$	$A_\oplus \simeq 0.4$	4.2×10^{-8}	3.4×10^{-8}	1.4×10^{-7}	3.0×10^{-6}
(9) Thermal emission	$\dfrac{4}{9}\dfrac{\mathscr{A}}{\mathscr{M}}\dfrac{\Phi_\odot}{c}\alpha\dfrac{\Delta\mathscr{T}}{\mathscr{T}_0}$	$\alpha = 0.4\text{--}0.7$ $\Delta\mathscr{T} = 1\text{--}20°$	9.5×10^{-8}	1.9×10^{-10}	2.7×10^{-10}	1.9×10^{-7}

Table 2.2 Uncertainties of the dynamical models.

Cause	Parameter	Relative uncertainty	Geosynchronous satellite	LAGEOS	Starlette	SEASAT
			Accelerations (cm s^{-2})			
(1) Earth's monopole	$GM_\oplus$	5×10^{-8}	1×10^{-6}	1×10^{-5}	4×10^{-5}	4×10^{-5}
(2) Earth's oblateness	$\bar{J}_{20}$	6×10^{-7}	4.5×10^{-10}	6×10^{-8}	5×10^{-7}	6×10^{-7}
(2) Geopotential harmonics						
$l = 2, m = 2$	$\bar{J}_{22}, \lambda_{22}$	10^{-3}	3×10^{-9}	6×10^{-7}	5×10^{-6}	5×10^{-6}
$l = 6, m = 6$	$\bar{J}_{66}, \lambda_{66}$	3×10^{-2}	2×10^{-11}	3×10^{-7}	2×10^{-5}	3×10^{-5}
$l = 18, m = 18$	$\bar{J}_{1818}, \lambda_{1818}$	1	negligible	7×10^{-10}	2×10^{-5}	4×10^{-4}
(3) Moon	$GM_{\mathbb{C}}$	10^{-5}	7×10^{-9}	2×10^{-9}	1×10^{-9}	1×10^{-9}
	$r_{\mathbb{C}}$	10^{-8}				
(3) Sun	$GM_\odot$	3×10^{-6}	1×10^{-9}	3×10^{-10}	2×10^{-10}	2×10^{-10}
	$r_\odot$	10^{-8}				
(6) Drag	ρ	$0.1–1$				
	C_D	$0.1–1$	O(?)	3×10^{-10}	7×10^{-9}	7×10^{-6}
	$\mathcal{A}$	$0.01–0.3$				
(7) Solar radiation pressure	$\mathcal{A}$	$0.01–0.3$				
	$\Phi_\odot$	2×10^{-3}	9×10^{-7}	6×10^{-9}	9×10^{-9}	2×10^{-6}
	Optical coefficients	$0.01–0.1$				
(8) Earth's albedo radiation pressure	$\mathcal{A}$	$0.01–0.3$				
	Optical coefficients	$0.01–0.1$	1×10^{-8}	8×10^{-9}	7×10^{-8}	1×10^{-6}
	$A_\oplus$	$0.1–1$				
(9) Thermal emission	α	$0.02–0.1$	1×10^{-7}	2×10^{-10}	3×10^{-10}	2×10^{-7}
	$\Delta \mathcal{T}$	$0.2–1$				

readily comparable in numerical work, one usually normalises the functions and includes their magnitude in the new 'normalised' coefficients $\bar{J}_{lm}$, which are used in the tables (for the normalising factors see Kaula 1966, equation (1.34)). To estimate the perturbing accelerations we have considered the angle-dependent part of the force, involving the normalised Legendre functions and their derivatives, to be of order unity (note that in some cases this assumption can underestimate the real force). The uncertainties shown in table 2.2 have been derived by comparison between the GEM-L2 geopotential model and the former GEM 9 model (see Lerch *et al* 1983).

(3) Gravitational perturbations due to other celestial bodies (the Moon, the Sun and the planets; $M_{\mathbb{C}}$, $M_{\odot}$ and M_{φ} are the masses of the Moon, the Sun and Venus respectively) are not caused by the full gravitational attraction of these objects but only by the corresponding 'tidal' terms, i.e. by the difference between the force on the Earth and that on the satellite. As we can see from table 2.2, the present accuracy of the ephemerides (giving the distances $r_{\odot}$, $r_{\mathbb{C}}$ and r_{φ} of the Sun, the Moon and Venus, respectively, from the Earth) is even higher than needed for satellite geodesy purposes, since for the Sun and the Moon the real problem is the mass ratio with respect to the Earth (see Ash *et al* 1971). Planetary perturbations are small—Venus provides the largest contribution—and cause no significant acceleration uncertainty on the Earth's satellites.

(4) This effect is due to the fact that the Earth's oblateness, whose (normalised) harmonic coefficient is $\bar{J}_{20}$, affects the motion of the Moon, and shifts the centre of mass of the Earth–Moon system with respect to the satellite. The corresponding effect due to the Moon's oblateness is smaller.

(5) The main correction to the Newtonian equations of motion introduced by general relativity is equivalent to an acceleration term given by the Earth's monopole term times the ratio between the Schwarzschild radius of the Earth ($GM_{\oplus}/c^2$, where c is the velocity of light) and the satellite's orbital distance (Ashby and Bertotti 1984).

(6) A prediction of the orbital perturbations resulting from drag is very complicated, since it would require the use of a model of the upper atmosphere, i.e. the density ρ of the resisting medium as a function of height. Even the best atmospheric models, however, are

inadequate to account for the large time variation of ρ, depending for instance on the level of solar and geomagnetic activity; in table 2.1 we assumed densities of 10^{-17} and 10^{-16} g cm^{-3} for the heights of Starlette and SEASAT respectively, which are typical values during a phase of intense solar activity. As a matter of fact, the procedure currently used is based on the determination from the tracking data of a few paramaters, like those appearing in the formula of table 2.1. In this formula C_D is a coefficient depending on the shape of the satellite and on the way the atmosphere molecules are re-emitted; $\mathscr{A}$ is the cross sectional area perpendicular to the line of motion, which is roughly constant for a nearly spherical satellite (like LAGEOS or Starlette), but variable and uncertain if the satellite's shape is more complex; and V is the velocity of the satellite relative to the atmosphere. As we will see in Chapter 6, besides a neutral gas drag, the charged particle drag can sometimes be relevant: for LAGEOS this latter effect (which depends in a complex way on the electromagnetic interaction between the satellite and the ionospheric plasma) is thought to be the dominant source of the observed deceleration of about 3×10^{-10} cm s^{-2} (Afonso *et al* 1980, Rubincam 1982). For Starlette and SEASAT the uncertainties of table 2.2 are derived by assuming reliabilities for the drag models of 10% and 30% respectively.

(7) The order of magnitude of the radiation pressure acceleration given in table 2.1 depends on the area-to-mass ratio of the satellite (with the complications quoted earlier when the cross section changes) and on the solar energy flux $\Phi_\odot$ (the so-called solar constant). Moreover, the precise intensity and direction of the force depend on the optical coefficients of the satellite's surface, yielding for each surface element the fraction of absorbed, reflected and diffused solar radiation (Anselmo *et al* 1981, 1983a). These coefficients can vary with time due to the interaction of the satellite with the environment of space. In table 2.2 we have assumed a 2% accuracy of the radiation pressure model for nearly spherical, geophysics-dedicated satellites like LAGEOS and Starlette and a 20% accuracy in the other cases. Finally, we have to remember that the modelling of the eclipse events, when the satellite moves in the Earth's shadow, is always very difficult and introduces further uncertainties. These problems will be discussed in Chapters 4 and 5.

(8) Earth-reflected solar radiation also produces a perturbing acceleration, proportional (for an order of magnitude estimate) to

ORDERS OF MAGNITUDE OF THE PERTURBING FORCES 19

the mean albedo of the Earth $A_\oplus$, and decreasing with the orbital distance according to the inverse square law. Note, however, that the optical properties of the Earth's surface are far from being uniform and time independent, because the albedo of different regions can range from about 0.2 to about 0.9, changing with time because of various meteorological and seasonal phenomena. This produces an additional uncertainty, which adds up to those discussed for the direct solar radiation. In table 2.2 we assumed a 25% uncertainty for the first two satellites, and a 50% uncertainty for the other two (the closer the satellites are to the Earth, the more relevant are the consequences of the albedo changes). This problem will be discussed in Chapter 5.

(9) The thermal re-emission of the radiation absorbed by the satellite (α is the absorption coefficient) occurs usually in an anisotropic way due to anisotropies of shape emissivity and surface temperature. The corresponding acceleration is proportional to $\Delta \mathcal{T}/\mathcal{T}_0$, the fractional difference of temperature between significant parts of the spacecraft. The values of α and $\Delta \mathcal{T}$ shown in table 2.1 refer respectively to a rapidly spinning 'passive' satellite (like LAGEOS and Starlette) or to a stabilised 'active' one, which has an internal energy conversion system (generally to emit radiowaves) and is equipped with solar panels. Obviously the availability of an accurate thermal model of the satellite can reduce the uncertainties of this term. Note that in cases of 'active' satellites, the emitted radiowave beams also produce a 'recoil' acceleration, whose order of magnitude is given by the same formula as the thermal emission, changing only 4/9 ($\Delta \mathcal{T}/\mathcal{T}_0$) into the (fractional) efficiency of the energy conversion system. This will also be discussed in Chapter 5.

Obviously the list in table 2.1 does not exhaust the sources of perturbation of satellite motion; we have only chosen phenomena producing effects significant with respect to the present accuracy of the observations. As a few examples of other existing physical mechanisms, we can quote the following (in parentheses we give an estimate of the corresponding acceleration on LAGEOS):

(a) micrometeorite drag ($\sim 10^{-11}$ cm s^{-2});
(b) Poynting–Robertson effect, due to the fact that even if the incident radiation is re-emitted isotropically, it is so only in the reference frame bound to the satellite (see Rubincam 1982) ($\sim 10^{-11}$ cm s^{-2});

(c) additional general relativistic perturbations (see Ashby and Bertotti 1984) ($\sim 10^{-12}$ cm s^{-2});
(d) emission of gravitational waves (see Peters 1964) ($\sim 10^{-35}$ cm s^{-2}).

The latter example is given to show that a physical mechanism, important in principle, is nevertheless completely negligible for artificial satellites.

2.3 TIDES AND APPARENT FORCES

Tides raised on the Earth by the the Moon and the Sun are relevant in satellite orbit determination in three ways:

(a) they cause periodic pulsations of the Earth, hence of the tracking stations (kinematic effect);
(b) they produce a variation in time of the geopotential, that affects the satellite orbit (dynamical effect);
(c) they perturb the rotation of the Earth, thus affecting the reference systems that have to be used in orbit computation (reference system effect).

In order to be able to compare such diverse effects either with each other or with the other perturbations discussed in §2.2, we transform all of them into equivalent accelerations and list them in table 2.3.

(1) The main tidal term produces a symmetric deformation of the Earth towards the perturbing body and opposite to it; therefore the period of the corresponding tidal wave is one half the synodic period of the perturbing body. Considering only the Moon, whose tidal effect on the Earth is about twice the effect of the Sun, a tracking station on the ground would oscillate (with period $\frac{1}{2}T_{\text{syn}}\mathbb{C} \sim 4.5 \times 10^4$ s and amplitude h) together with the solid body of the Earth. If this effect were totally ignored in the satellite orbit determination, the station acceleration would appear as a residual acceleration on the satellite. The amplitude h depends upon the elastic response of the Earth to the perturbing potential; the relevant response coefficient, or Love number, is usually denoted h_2, and its value is fairly well known, giving uncertainties of the order of one centimetre or less in h (see Lambeck 1980).

(2) Tidal waves raised in the oceans also affect the station sites'

Table 2.3 Equivalent accelerations of tidal effects.

Cause	Formula	Parameters (in CGS units)	Relative uncertainty	Accelerations (cm s^{-2})				
				Geosynchronous satellite $a = 42\,160$ km	LAGEOS 12 270	Starlette ~7300	SEASAT (or ERS-1) ~7100
(1) Kinematic solid tide	$h\left(\dfrac{2\pi}{T_{\text{sym}\mathbb{C}/2}}\right)^2$	$h \simeq 30$ cm	$\gtrsim 0.03$	5.8×10^{-7}	5.8×10^{-7}	5.8×10^{-7}	5.8×10^{-7}
(2) Kinematic oceanic loading	$h_L\left(\dfrac{2\pi}{T_{\text{sym}\mathbb{C}/2}}\right)^2$	$h_L \simeq 5$ cm	0.2	10^{-7}	10^{-7}	10^{-7}	10^{-7}
(3) Dynamic solid tide	$3k_2\,\dfrac{GM_{\mathbb{C}}}{r_{\mathbb{C}}}\left(\dfrac{R_\oplus}{r_{\mathbb{C}}}\right)^2 \dfrac{R_\oplus^3}{r^4}$	$k_2 \simeq 0.3$	0.01	2.7×10^{-8}	3.7×10^{-6}	3×10^{-5}	3.3×10^{-5}
(4) Dynamic oceanic tide	~0.1 of the dynamic solid tide		0.1	2.7×10^{-9}	3.7×10^{-7}	3×10^{-6}	3.3×10^{-6}
(5) Reference system: non-rigid Earth nutation (fortnightly term)	0.002 arcsec in 14 days	Nutation coefficient	0.1	1.2×10^{-9}	3.5×10^{-10}	2×10^{-10}	2×10^{-10}

positions. Displacements of nearby water masses produce a time-dependent elastic response of the continental shelves (oceanic loading) and this changes the station position with respect to the theoretical tidal displacement corresponding to an oceanless solid Earth. Of course the size of this effect depends strongly on the geographical situation of the station, e.g. the corresponding displacements h_L can be as large as 12 cm on a peninsula protruding into an ocean (such as Cornwall) and less than 1 cm in the middle of a continental plate. Even if some modelling is possible, such an effect is determined essentially from ground-based tidal measurements, leaving uncertainties of the order of one centimetre. The equivalent acceleration of the kinematical effects and their uncertainties, as listed in table 2.3, may appear large with respect to the other forces listed in tables 2.1 and 2.2; however, they affect the satellite tracking data with well defined frequencies, i.e. the tidal frequencies themselves, and do not have secular or long-periodic effects. It is therefore possible to separate the kinematical effects from other dynamical effects in such a way that their uncertainty does not affect the uncertainty of other geophysically relevant parameters. Of course, when station positioning is the main goal the tidal displacements must be properly modelled.

(3) The main term of the tidal perturbation to the geopotential is obtained as the corresponding (i.e. with the same frequency) term of the perturbing gravitational potential generated by the other body, say the Moon, times a Love number k, measuring the ratio between the response of the real Earth and the theoretical response of a perfect fluid. Since the main term is a quadrupole one (symmetric elongation of the Earth in the direction of the perturbing body), the corresponding acceleration felt by the satellite is proportional to r^{-4}. The main Love number k_2 is currently determined from geodetic satellite orbits with a fairly good accuracy, because it produces long-term effects (Lerch *et al* 1983). An additional source of model errors is the truncation of the tidal potential series expansion to a finite (and usually very small) number of terms.

(4) Tides appear in the oceans as waves that move on the surface forced by the gravitational perturbations of the Moon and the Sun. However, waves cannot propagate in a frictionless way, but do dissipate energy and also interact with the sea floor (especially in shallow waters) and with the shore line. As a result, tidal waves on the seas are generally not in phase with the perturbing potential.

TIDES AND APPARENT FORCES

The water masses displaced in such waves also give a contribution to the variation of the geopotential with time. Therefore this effect must be considered in modelling the satellite orbit together with the solid tide; since it is very difficult to separate the two effects on the orbits, the uncertainty in the difficult modelling of oceanic tides limits the accuracy in the modelling of the solid tides as well.

(5) Since the Earth is not rigid, it does not respond to external torques as a rigid body should, and its rotation axis undergoes forced nutations that are different from the theoretical nutation described by the rigid-Earth nutation tables (Woolard theory). The most important difference lies in the fortnightly term (with argument twice the mean anomaly of the Moon) and amounts to two milliarcseconds. Were this effect ignored, the satellites would appear to oscillate by that amount as a result of an incorrect definition of the reference system in which the equations of motion are written. Tables are now available modelling the elastic response of the Earth according to up to date Earth models (Wahr 1981); even if it is not easy to assess how accurate these new models are, we can assume that an improvement of one order of magnitude in the accuracy of nutation tables has been achieved.

There are uncertainties in the determination of the reference frames used in orbit computation that are not due to tides; these depend upon the unpredictable variations of the Earth's rotation. Even if this represents a completely different physical phenomenon, for comparison purposes we discuss their effects in terms of apparent forces that result from the computation being performed in a wobbling frame.

The angular velocity vector $\omega_\oplus$ of Earth's rotation is known to undergo variations both in direction ('wobble') and in magnitude (variation of the length of the day (LOD), or Δ(LOD)) (see Munk and MacDonald 1960, Lambeck 1980). For polar wobble the two main periods are 1 year and about 14 months (Chandler wobble); for the LOD the main component is annual too. In an inertial reference frame the effects on the range data allowing orbit determination for artificial satellites are kinematical, since the tracking stations are bound to the Earth. In an Earth-fixed reference frame apparent forces arise due to the fact that the Earth rotates with the angular velocity

$$\omega_\oplus(t) = \omega_\oplus^0 + \Delta\omega_\oplus(t) \qquad (2.2)$$

where $\omega_\oplus^0$ is the constant nominal angular velocity of the Earth directed along the z axis ($|\omega_\oplus^0| = 7.29 \times 10^{-5}$ rad s^{-1}) while $\Delta\omega_\oplus(t)$ is a small time-dependent vector accounting for polar wobble and variations of LOD. The perturbing acceleration in an Earth-fixed reference frame is (Farinella *et al* 1981):

$$F = P \times \Delta\dot{\omega}_\oplus + 2\dot{P} \times \Delta\omega_\oplus + 2(\omega_\oplus^0 \cdot \Delta\omega_\oplus)P - (\Delta\omega_\oplus \cdot P)\omega_\oplus^0$$
$$- (\omega_\oplus^0 \cdot P)\Delta\omega_\oplus \quad (2.3)$$

where P is the position vector of the satellite in the Earth-fixed frame and $\dot{P}$ the corresponding velocity vector; terms that are of second order in $|\Delta\omega_\oplus|$ have been omitted. Each of the five terms of equation (2.3) has an effect on the satellite with its characteristic signature, that is different for a $\Delta\omega_\oplus$ parallel to $\omega_\oplus^0$ (i.e. due to Δ(LOD)) or perpendicular to it (i.e. due to polar wandering). We list an order of magnitude computation of all these apparent forces in table 2.4, computed only for a typical geosynchronous satellite (in almost circular, equatorial orbit) and for LAGEOS (low eccentricity but high inclination). For lower satellites the perturbations caused by the anomalies in the gravity field of the Earth of high degree and order are much more important; they are not suitable for the determination of polar motion. Table 2.4 has been compiled assuming a variation in the pole's position of about $0''.2$ over a time-scale of about one year and an annual variation in the length of the day of amplitude 10^{-3} s. This means:

$$\Delta\omega_{\oplus x} \simeq \Delta\omega_{\oplus y} \simeq \frac{0.2}{2 \times 10^5} \omega_\oplus^0 \simeq 7.3 \times 10^{-11} \text{ rad s}^{-1}$$

$$\Delta\omega_{\oplus z} \simeq \frac{10^{-3}}{8.6 \times 10^4} \omega_\oplus^0 \simeq 8.5 \times 10^{-13} \text{ rad s}^{-1}$$
$$\quad (2.4)$$
$$\Delta\dot{\omega}_{\oplus x} \simeq \Delta\dot{\omega}_{\oplus y} \simeq \Delta\omega_{\oplus x} \frac{2\pi}{3 \times 10^7 \text{ s}} \simeq 1.5 \times 10^{-17} \text{ rad s}^{-2}$$

$$\Delta\dot{\omega}_{\oplus z} \simeq \Delta\omega_{\oplus z} \frac{2\pi}{3 \times 10^7 \text{ s}} \simeq 1.7 \times 10^{-19} \text{ rad s}^{-2}.$$

The uncertainty in the knowledge of the pole position has recently been reduced by analysing the orbit of LAGEOS (Lerch *et al* 1983) and it now amounts to a few milliarcseconds, at least for variations with a period longer than five days (relative uncertainty

Table 2.4 Apparent accelerations due to polar wobble and length of the day variation.

Term	Geosynchronous satellite $a = 42\,160$ km, $e \simeq 10^{-3}$, $I \simeq 1°$		LAGEOS $a = 12\,270$ km, $e \simeq 4 \times 10^{-3}$, $I = 109°$	
	Accelerations due to polar wobble (cm s^{-2})	Accelerations due to Δ(LOD)(cm s^{-2})	Accelerations due to polar wobble (cm s^{-2})	Accelerations due to Δ(LOD)(cm s^{-2})
$(\Delta\omega_\oplus \cdot P)\omega_\oplus^0$	2.2×10^{-5}	4.5×10^{-9}	2.0×10^{-6}	7.0×10^{-8}
$2\dot{P} \times \Delta\omega_\oplus$	7.8×10^{-7}	5.0×10^{-10}	7.7×10^{-5}	2.0×10^{-7}
$P \times \Delta\dot{\omega}_\oplus$	1.0×10^{-9}	7.5×10^{-10}	1.7×10^{-8}	6.8×10^{-11}
$2(\omega_\oplus^0 \cdot \Delta\omega_\oplus)P$		5.2×10^{-7}		1.5×10^{-7}
$(\omega_\oplus^0 \cdot P)\Delta\omega_\oplus$	3.9×10^{-7}	4.5×10^{-9}	6.2×10^{-6}	7.2×10^{-8}

1%). The relative uncertainty in $\Delta(\text{LOD})$ also amounts to a few per cent (Lerch *et al* 1983). Therefore, if the data obtained in this way are used as an *a priori* model, uncertainties become quite small for all but few terms.

3

TOOLS FROM CELESTIAL MECHANICS

This chapter reviews some basic theory of satellite orbits, including the unperturbed Keplerian orbits (§3.1), the effects of perturbing forces on the Keplerian elements (§3.2), the non-singular orbital elements (§3.3) and the method of averaging over entire orbits to compute the long-term perturbations (§3.4). Any reader familiar with celestial mechanics can omit this chapter, or at least parts of it. For this chapter we have used many textbooks, looking for the simplest presentations; although no specific references are given in the text, we are particularly indebted to King-Hele (1964), Sterne (1960), Burns (1976) and Roy (1978).

3.1 KEPLERIAN ORBITS

We first discuss the unperturbed orbit, that is the orbit a satellite would follow if the Earth were a perfect sphere and no other gravitational or non-gravitational perturbations were acting upon it. The differential equations for the geocentric position vector r of the spacecraft would be the ones of the central force problem (dots indicate derivation with respect to time):

$$\ddot{r} = -\mu r/r^3 \qquad (3.1)$$

where $\mu = GM_\oplus = 3.986\,006 \times 10^{20}$ cm^3 s^{-2} (according to the last determination, the GEM-L2 geopotential model (Lerch *et al* 1983)). The problem (3.1) is integrable, that is there are sufficient time-independent first integrals (i.e. functions constant along the motion) to specify each orbit—these are the angular momentum integral

$$H = r \times \dot{r} \qquad \dot{H} = 0 \qquad (3.2)$$

and the Lenz vector

$$e = \dot{r} \times H/\mu - r/r \qquad \dot{e} = 0. \qquad (3.3)$$

The vector integrals H and e provide five independent constants (not six, because $H \cdot e = 0$) and are enough to specify the orbit that will lie in the plane perpendicular to H and have a shape determined by e. The angle between e and r (as seen from H) is the true anomaly f

$$\cos f = r \cdot e/(er). \qquad (3.4)$$

By using (3.3) and (3.4) the radius r can be expressed as a function of the true anomaly f and of the scalar integrals e (eccentricity) and H (angular momentum, scalar)

$$r = \frac{H^2/\mu}{1 + e \cos f} \qquad (3.5)$$

provided $H \neq 0$. Equation (3.5) describes in polar coordinates (r, f) a conic section: we will be concerned only with either circular ($e = 0$) or elliptic ($0 < e < 1$) orbits. These are the only ones to be limited, with distance r in the range

$$r_1 = \frac{H^2}{\mu(1 + e)} \leqslant r \leqslant \frac{H^2}{\mu(1 - e)} = r_2 \qquad (3.6)$$

and the major axis of the ellipse will therefore be

$$2a = r_1 + r_2 = \frac{2H^2}{\mu(1 - e^2)}. \qquad (3.7)$$

To completely specify the orbit we can use a set of six quantities, the Keplerian elements: the semi-major axis a, the eccentricity e and the true anomaly at a specified time $f(t)$ define the orbit in the reference frame defined by the direction of the perigee (along e) as x axis and the normal to the orbital plane (along H) as z axis. In this frame the coordinates of the satellite are: $(r \cos f, r \sin f, 0)$, with r given by (3.5).

KEPLERIAN ORBITS

Three additional elements are required to specify the orientation of the frame defined by e, H with respect to the standard orthonormal reference frame x, y, z. The latter is in principle arbitrary for a purely central force problem such as (3.1); however, for artificial Earth satellites it is convenient to have z along the polar axis and x, y in the equatorial plane. (We will however assume that x, y, z is an inertial frame, while the polar axis does change; the choice of the appropriate reference system is one of the most sensitive problems in satellite geodesy that we will not address here—see §2.3.) Once x, y, z have been chosen, we define the unit vector Q as the direction of the ascending node. That is, Q must lie in the intersection of the orbital plane (perpendicular to H) with the 'equatorial' x, y plane, and be oriented toward the point in which the orbit crosses the equatorial plane going from below it to above it as time goes by:

$$Q = z \times H / |z \times H| \qquad (3.8)$$

(see figure 3.1).

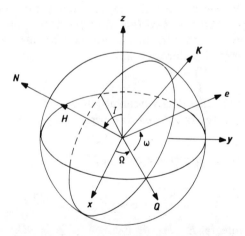

Figure 3.1

The longitude of the ascending node Ω is the angle between the x axis and the ascending node Q, as seen from z:

$$Q = x \cos \Omega + y \sin \Omega. \qquad (3.9)$$

The inclination I is the angle between z and H, as seen from Q:

$$\cos I = z \cdot H/H \tag{3.10}$$

that is, if k is a unit vector in the orbital plane perpendicular to Q (and chosen in such a way that the angle between Q and k is $+\pi/2$ as seen from H), then

$$z = (H \cos I)/H + k \sin I. \tag{3.11}$$

However, I is assumed to be $0 \leqslant I \leqslant \pi$. The argument of perigee ω is the angle between Q and e, as seen from H:

$$e/e = Q \cos \omega + k \sin \omega. \tag{3.12}$$

In this way a, e, I, Ω, ω specify the orbit (without the parametrisation by the time) as well as e, H.

Two more integrals are available, but they are functions of the previous ones. The area integral, that is the time derivative of the area A swept by the radius vector r in the orbit plane:

$$\dot{A} = r^2 \dot{f}/2 = H/2 \tag{3.13}$$

turns out to be half of the scalar angular momentum. From (3.13) we can deduce *Kepler's third law:* the period P of an elliptic (or circular) orbit (or better the mean motion n given by $nP = 2\pi$) is related to the semi-major axis only:

$$n^2 a^3 = \mu. \tag{3.14}$$

Thanks to (3.13) the velocity v can also be specified in terms of the true anomaly f: the radial and transverse components v_R, v_T of v are

$$v_T = H/r = (1 + e \cos f)\mu/H \tag{3.15}$$

$$v_R = \dot{r} = \frac{H^2}{\mu} \frac{e \sin f}{(1 + e \cos f)^2} \dot{f} = \frac{\mu}{H} e \sin f. \tag{3.16}$$

The energy integral E

$$E = v^2/2 - \mu/r \tag{3.17}$$

turns out to be related to e and H by the formula

$$e^2 = 1 + 2EH^2/\mu^2. \tag{3.18}$$

However, E is especially significant because the semi-major axis a

KEPLERIAN ORBITS

is a function of E only:

$$a = -\mu/(2E). \quad (3.19)$$

Another useful formula is derived from (3.18) and (3.19):

$$H^2 = \mu a(1 - e^2). \quad (3.20)$$

The main problem about Keplerian elliptic orbits we have still to address is how to compute the relation between the position along the orbit and the time, that is to solve the differential equation $\dot{f} = H/r^2$. The latter does not have an explicit solution in terms of elementary analytic functions; therefore we are forced to resort to an indirect, implicit solution. We first define the eccentric anomaly u either geometrically, as in figure 3.2, or analytically through the relation with the true anomaly and the position of the satellite in the orbital plane

$$r \cos f = a (\cos u - e)$$
$$r \sin f = a\beta \sin u \qquad \beta = \sqrt{1 - e^2}. \quad (3.21)$$

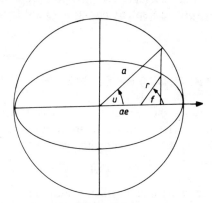

Figure 3.2

The radius r is easily expressed as a function of the eccentric anomaly u:

$$r = a(1 - e \cos u). \quad (3.22)$$

Other useful relationships between true and eccentric anomaly are deduced from (3.21), (3.22):

$$\cos u = \frac{\cos f + e}{1 + e \cos f} \qquad (3.23)$$

$$\sin f = \frac{\beta \sin u}{1 - e \cos u}. \qquad (3.24)$$

The differential equation controlling the evolution of u with time can be deduced by taking the time derivative of (3.22) and by using (3.16) and (3.24):

$$\dot{u} = na/r = n/(1 - e \cos u) \qquad (3.25)$$

and by integration:

$$u - e \sin u = n(t - t_0) = M \qquad (3.26)$$

where t_0 is a constant of integration specifying the time of a passage through $u = 0$ (that is also $f = 0$, a pericenter passage) and the angle M is the mean anomaly. M is defined by the requirement that its time derivative is the mean motion n and its value is a multiple of 2π at each pericenter passage. Thus the position along the orbit can be specified by solving Kepler's equation (3.26) for u and by using (3.21).

3.2 VARIATIONS OF THE ELEMENTS

We will now discuss the changes in the orbital elements produced by the action of a perturbing acceleration F; that is when the equations of motion in cartesian coordinates are

$$\ddot{r} = -\mu r/r^3 + F \qquad (3.27)$$

F being any known function of r, $\dot{r}$ and time t. The main idea underlying the perturbative equations we are going to discuss is the definition of osculating orbital elements (or even osculating energy, angular momentum, etc). Given an orbit $r(t)$, $\dot{r}(t)$, for every time t_1 we can compute the Keplerian elements a, e, I, Ω, ω and $f(t_1)$ (or $M(t_1)$) of a Keplerian orbit with initial conditions $r(t_1)$, $\dot{r}(t_1)$. They give the orbit the satellite would follow if at time t_1 the perturbation F were suddenly turned off.

These osculating elements are functions of the time t_1: as a result, even the elements that are constant in a Keplerian orbit, such as the semi-major axis a, are now time-varying. Our purpose is to write down the ordinary differential equations that describe their time dependence. In the same way we can compute $\dot{E}$, that is not the time derivative of the total energy of the system — which is in many cases conserved — but the time derivative, along the perturbed orbit, of the function $E(r, \dot{r})$ defined by formula (3.17). The tricky point in the computations that follow is that the formulae connecting the Keplerian elements, the integrals of the unperturbed orbits such as E, H, e and the instantaneous position and velocity r, $\dot{r}$ still hold for osculating elements and integrals and for the perturbed orbits: e.g. (3.5), (3.15) and (3.16). On the contrary, whenever, as a result of a differentiation with respect to time, quantities such as $\dot{E}$, $\dot{H}$, $\dot{a}$, etc, appear, these refer to the time variation of the osculating elements and therefore will contain the perturbing force F. Some additional problems arise in computing the time derivatives of the anomalies (f, u or M) that are not constant even in the unperturbed orbit.

The effects of the perturbing acceleration F can be very different according to its direction F/F and to its relationship with the orbital geometry. Let us decompose F in three orthogonal components, acting in the radial direction (along r), in the out-of-orbital-plane direction (along H), and transversely in the orbital plane (along $H \times r$). This will give the perturbative equations in the Gauss form. Let us call the projections of F along these three directions R, W and T respectively:

$$R = F \cdot r/r \qquad W = F \cdot H/H \qquad T = F \cdot (H \times r)/Hr \qquad (3.28)$$

where H is the osculating angular momentum, given by (3.2). The osculating energy changes at a rate given by the power $\dot{E} = F \cdot \dot{r}$ and $\dot{r}$ is expressed as in (3.15) and (3.16):

$$\dot{E} = Rv_R + Tv_T = [R(e \sin f) + T(1 + e \cos f)]\mu/H. \quad (3.29)$$

By time differentiation of equation (3.19) we get $\dot{a} = 2a^2\dot{E}/\mu$, and the first equation of the 'variation of elements' is so obtained:

$$\dot{a} = \frac{2}{n\beta} [T + e(T \cos f + R \sin f)] \qquad (3.30)$$

with $\beta = (1 - e^2)^{1/2}$. Equation (3.30) makes it easy to appreciate that the transverse perturbation T is the most important in changing the size of the orbit (hence the period), especially for low eccentricity e.

To find the equations for $\dot{e}$, we use the fact that the angular momentum scalar H is changed only by a torque along the same direction of H, that is $\dot{H} = rT$. On the other hand by differentiating (3.20)

$$rT = \dot{H} = [\mu\dot{a}(1 - e^2) - 2e\dot{e}\mu a]/(2H) \qquad (3.31)$$

and we can solve for $\dot{e}$ in (3.31) by substituting (3.22) for r and (3.30) for $\dot{a}$:

$$\dot{e} = \frac{\beta}{na} [R \sin f + T (\cos f + \cos u)]. \qquad (3.32)$$

The changes of the orientation of the osculating orbital plane, that is of the elements I and Ω, can be computed by taking the time derivative of the unit vector N perpendicular to the orbital plane

$$H = HN \qquad \dot{H} = \dot{H}N + H\dot{N}. \qquad (3.33)$$

The change in the angular momentum is due to the torque generated by F

$$\dot{H} = r \times \ddot{r} = r \times F \qquad (3.34)$$

but the component T of F will change H parallel to itself as before: only W will be effective in turning the normal N

$$\dot{N} = (r \times N)W/H. \qquad (3.35)$$

The change $\dot{N}$ can be decomposed into a change along the line of the nodes, that is $\dot{N} \cdot Q$, and a change perpendicular to Q in the orbit plane, that is $\dot{N} \cdot k$ (where k, as in (3.11), is $k = N \times Q$); $N \cdot \dot{N} = 0$ because $N \cdot N = 1$ always. Then

$$\dot{N} \cdot Q = \frac{W}{H} (r \times N) \cdot Q = \frac{W}{H} (N \times Q) \cdot r = \frac{W}{H} k \cdot r \qquad (3.36)$$

$$\dot{N} \cdot k = \frac{W}{H} (r \times N) \cdot k = \frac{W}{H} (N \times k) \cdot r = -\frac{W}{H} Q \cdot r. \qquad (3.37)$$

The only computational trick being used is the transformation of the triple products $(A \times B) \cdot C = (B \times C) \cdot A$. The components of r along Q, k can be computed from the components $r \cos f$, $r \sin f$ along e and along the perpendicular to e and from the definition of ω (3.12):

$$r \cdot Q = r \cos(\omega + f) \qquad r \cdot k = r \sin(\omega + f). \qquad (3.38)$$

VARIATIONS OF THE ELEMENTS

On the other hand, the components of $\dot{N}$ can be interpreted as follows: from the definition (3.10) of the inclination I, that is $\cos I = N \cdot z$, by taking the time derivative and by using (3.11)

$$-\dot{I} \sin I = \dot{N} \cdot z = \dot{N} \cdot (N \cos I + k \sin I) = \sin I\, (\dot{N} \cdot k); \quad (3.39)$$

and by substituting into (3.37), using also (3.38)

$$\dot{I} = \frac{W}{H} r \cos(\omega + f). \quad (3.40)$$

In a similar way, from the definition of Ω (3.9), by taking the time derivative

$$\dot{Q} = \dot{\Omega}[-x \sin \Omega + y \cos \Omega] \quad (3.41)$$

since $N \cdot Q = 0$, $N \cdot \dot{Q} = -\dot{N} \cdot Q$ and by scalar product of (3.41) by N we get

$$-\dot{N} \cdot Q = \dot{\Omega}[-x \sin \Omega + y \cos \Omega] \cdot N. \quad (3.42)$$

The vector between square brackets in (3.42) is a unit vector in the x, y plane orthogonal to Q, thus the angle it forms with N is the angle between N and the plane x, y, that is $\pi/2 - I$, and the scalar product is $\sin I$. By (3.36) and (3.38)

$$\dot{\Omega} \sin I = \frac{W}{H} r \sin(\omega + f). \quad (3.43)$$

The computation of $\dot{\omega}$ is slightly more difficult because ω is an angle measured in the orbital plane, but is changed also by the changes in the orientation of the orbital planes, hence $\dot{\omega}$ contains W as well as R and T — this is because it is measured from the node. The simplest way out is to compute the combination $\dot{\omega} + \dot{\Omega} \cos I$, that is an angular velocity around the axis of the angular momentum H. The size of the latter is still $H = rv_T$. However, v_T is not $r\dot{f}$ (as it would be in the unperturbed motion) because $\dot{f}$ is now the time derivative of the osculating f along the perturbed orbit and not the time derivative of the true anomaly of an unperturbed orbit. The transverse velocity contains $\dot{\omega} + \dot{\Omega} \cos I$ as well:

$$H/r^2 - \dot{f} = \dot{\omega} + \dot{\Omega} \cos I. \quad (3.44)$$

On the contrary the formulae not containing $\dot{f}$ are the same in the unperturbed and the perturbed case. For example $H^2 = \mu r(1 + e \cos f)$ and the time derivative is

$$2 H\dot{H} = \mu \dot{r}(1 + e \cos f) + \mu r \, \dot{e} \cos f - \mu r e \, \dot{f} \sin f. \quad (3.45)$$

We now substitute into (3.45): $\dot{H} = rT$; (3.16) for $\dot{r}$; (3.32) for $\dot{e}$; (3.44) for $H/r^2 - \dot{f}$; and get

$$\dot{\omega} + \dot{\Omega} \cos I = \frac{H}{\mu e} \left(-R \cos f + T \frac{2 - \cos^2 f - \cos u \cos f}{\sin f} \right). \quad (3.46)$$

The expression multiplying T can be simplified by using (3.23) for $\cos u$ and (3.24) for $\sin f$:

$$\dot{\omega} + \dot{\Omega} \cos I = \frac{\beta}{ena} \left[-R \cos f + T \left(\sin f + \frac{1}{\beta} \sin u \right) \right]. \quad (3.47)$$

The last equation for the variation of the elements should give the time variation of some osculating anomaly, say the mean anomaly M. However, $\dot{M}$ does contain both the mean motion n of the osculating ellipse and the effect of the perturbation. In order to find an element that changes 'slowly', that is with a time derivative going to zero as $F \to 0$, we define a new variable η by

$$M(t) = \eta + \rho \qquad \rho = \int_{\bar{t}}^{t} n(t') \, dt' \quad (3.48)$$

where $M(t)$ is the osculating mean anomaly and $\bar{t}$ is an arbitrary origin in time. By differentiating (3.48) and Kepler's equation (3.26) we get

$$\dot{\eta} = \dot{M} - n = (1 - e \cos u)\dot{u} - \dot{e} \sin u - n. \quad (3.49)$$

To compute $\dot{u}$ along the perturbed orbit we use (3.22), differentiate with respect to time and use (3.16)

$$\dot{a}r/a - \dot{e}a \cos u + ea\dot{u} \sin u = \dot{r} = nae \sin f/\beta. \quad (3.50)$$

By substituting (3.30) for $\dot{a}$ and (3.32) for $\dot{e}$ and after some reduction (with the use of (3.23) and (3.24)) we find

$$n - \frac{r}{a} \dot{u} = -R \left[\frac{2r}{na^2} - \frac{\cos u}{ae\beta^2} \right] - T \left[\frac{\beta \sin u}{nae} \left(\frac{a}{r} - 1 \right) \right] \quad (3.51)$$

VARIATIONS OF THE ELEMENTS

and by substituting (3.51) and (3.32) into (3.49) we find

$$\dot{\eta} = R\left[\frac{\cos u}{nae\beta^2} - \frac{\beta \sin f \sin u}{na} + \frac{2r}{na^2}\right]$$
$$+ T\left[\frac{\beta \sin u}{nae}\left(\frac{a}{r} - 1\right) - \frac{\beta}{na}\sin u(\cos f + \cos u)\right]. \quad (3.52)$$

However, the most important perturbative effect 'along track', that is in the osculating mean anomaly, is not due to $\dot{\eta}$ but to $\dot{n}$: this is because the changes in the semi-major axis will imply a change in mean motion. From (3.14) we get

$$n = \mu^{1/2} a^{-3/2} \qquad \dot{n} = -\frac{3}{2}\frac{n}{a}\dot{a}. \quad (3.53)$$

If we decompose $M = \eta + \rho$, the equation for ρ will be of the second order. By (3.53) and (3.30) we obtain

$$\ddot{\rho} = \dot{n} = -\frac{3}{a\beta}[T + e(T\cos f + R\sin f)]. \quad (3.54)$$

If T is such that there is a constant drift in a and n, e.g. if T is constant, the changes in ρ will accumulate quadratically with time and will therefore be much larger than the change in η. If on the contrary F is such that a and n oscillate around an average value, the two equations (3.54) and (3.52) will be of the same relevance. This will be an important point in many discussions in the following chapters.

For perturbing accelerations that act in the direction of the velocity vector v (such as the main component of the drag) it is better to decompose F in the tangential component $F_v = \boldsymbol{F} \cdot \boldsymbol{v}/v$ and the normal component F_N (along the inward normal to the orbit in the osculating orbital plane). Then

$$T = F_v \frac{v_T}{v} + F_N \frac{v_R}{v} \qquad R = F_v \frac{v_R}{v} - F_N \frac{v_T}{v}. \quad (3.55)$$

By using (3.15) and (3.16) and substituting into (3.30), (3.32) and (3.47) the variations can be given as functions of F_v, F_N. The most important are

$$\dot{a} = F_v 2a^2 v/\mu \quad (3.56)$$

which of course could be deduced directly from $\dot{E} = \boldsymbol{F} \cdot \dot{\boldsymbol{r}} = vF_v$, and

$$\dot{e} = [2F_v(e + \cos f) - (a/r)F_N \sin f]/v. \quad (3.57)$$

3.3 NON-SINGULAR ELEMENTS

The equations for the variation of the elements ω, M and Ω are singular for $e \to 0$ and/or for $I \to 0$. However, there is no physical instability corresponding to this mathematical singularity: e.g. for e very small, a large change in ω corresponds to a small change in the position and velocity vectors r, $\dot{r}$ and can thus be produced by a small acceleration F. To avoid the terms of the form $1/e$ or $1/\sin I$ in the equations we need to remove the singularity from the coordinate change between r, $\dot{r}$ and the set of six elements.

To remove the singularity for $e = 0$ we define three new elements to replace e, ω and M:

$$h = e \sin \omega \qquad k = e \cos \omega \tag{3.58}$$

which are in the same relationship with e, ω as cartesian coordinates are to polar ones; and

$$\lambda = \omega + M \tag{3.59}$$

which is the mean argument of latitude, an angle measured from the node in the orbital plane. For $e = 0$, λ is defined as the angle between the radius vector and the node vector Q, and the two definitions match in a smooth way. The equations for the variation of h and k can be obtained by

$$\dot{h} = \dot{e} \sin \omega + e\dot{\omega} \cos \omega \qquad \dot{k} = \dot{e} \cos \omega - e\dot{\omega} \sin \omega \tag{3.60}$$

and by (3.32), (3.47) and (3.43). As a result $\dot{h}$ and $\dot{k}$ are defined even for $e = 0$ without singularities.

The purpose of the use of h and k is twofold: it avoids the appearance of very large variations of the elements for small e, but can also be used to provide approximate equations by truncation of the terms containing $e^2 = h^2 + k^2$ or higher order in e. When the substitutions in (3.60) are done for the latter purpose, we obtain

$$na\dot{h} = -R \cos(\omega + f) + T \sin(\omega + f) + T \sin(\omega + u)$$
$$+ kW \cot I \sin(\omega + f) + O(e^2)$$
$$\tag{3.61}$$
$$na\dot{k} = R \sin(\omega + f) + T \cos(\omega + f) + T \cos(\omega + u)$$
$$+ hW \cot I \sin(\omega + f) + O(e^2).$$

where the symbol $O(\ldots)$ indicates the order of the omitted terms.

NON-SINGULAR ELEMENTS

Further simplifications to (3.61) can be done by exploiting the expansion of cos f, sin f, cos u, sin u in D'Alembert series, that is in Fourier series in the mean anomaly M with coefficients containing higher powers of the eccentricities for the higher harmonics (see Kovalevsky 1963, Ch. 4):

$$\begin{aligned}\cos f &= \cos M - e + e \cos 2M + O(e^2) \\ \sin f &= \sin M + e \sin 2M + O(e^2)\end{aligned} \quad (3.62)$$

$$\begin{aligned}\cos u &= \cos M - \tfrac{1}{2} e + \tfrac{1}{2} e \cos 2M + O(e^2) \\ \sin u &= \sin M + \tfrac{1}{2} e \sin 2M + O(e^2)\end{aligned} \quad (3.63)$$

and by simple use of the addition formulae for sines and cosines:

$$na\dot{h} = -R \cos \lambda + 2T \sin \lambda - 3Th/2 - Rk + 3T(k \sin 2\lambda - h \cos 2\lambda) \\ - R(k \cos 2\lambda + h \sin 2\lambda) - kW \cot I \sin \lambda + O(h^2 + k^2) \quad (3.64a)$$

and

$$na\dot{k} = R \sin \lambda + 2T \cos \lambda - 3T k/2 + Rh + 3T(k \cos 2\lambda + h \sin 2\lambda) \\ + R(k \sin 2\lambda - h \cos 2\lambda) + hW \cot I \sin \lambda + O(h^2 + k^2). \quad (3.64b)$$

For $\dot{\lambda}$ the same problems as for $\dot{M}$ occur; we define a new variable ε by:

$$\varepsilon + \int_{\bar{t}}^{t} n(t') \, dt' = \lambda = M + \omega \qquad \varepsilon = \eta + \omega \quad (3.65)$$

and the change in ε can be computed by

$$\dot{\varepsilon} = \dot{\eta} + \dot{\omega}. \quad (3.66)$$

By using (3.52), (3.47) and (3.43), after awful computations, the $1/e$ terms cancel out and we get

$$\dot{\varepsilon} = \frac{R}{na} (-2 + \tfrac{3}{2} e \cos f) + \frac{T}{na} \tfrac{1}{2} e \sin f \\ - \frac{W}{na} \cot I (1 - e \cos f) \sin (\omega + f) + O(e^2). \quad (3.67)$$

In (3.67) the terms with cos f, sin f are non-singular because they contain a factor e and can therefore be expressed as a combination of λ, h, k.

For most of today's geodetic satellites the eccentricity is small and the inclination large (e.g. LAGEOS, GPS) and the formulae regularised for $e = 0$ are convenient. However, there is at least one important class of satellites with small e but also small I: the geostationary satellites. For these a set of variables non-singular for both $e = 0$ and $I = 0$ has to be used, together with the longitude of the perigee:

$$\tilde{\omega} = \omega + \Omega \tag{3.68}$$

where for $I = 0$, $\tilde{\omega}$ is the angle between x and e. We redefine h, k as

$$h = e \sin \tilde{\omega} \qquad k = e \cos \tilde{\omega} \tag{3.69}$$

and a similar procedure is used for I, Ω, by defining

$$p = \tan I \sin \Omega \qquad q = \tan I \cos \Omega. \tag{3.70}$$

The regularised set of elements is completed by redefining λ as the mean longitude

$$\lambda = \Omega + \omega + M = \tilde{\omega} + M. \tag{3.71}$$

The equations for the variations of h and k can be computed from a formula similar to (3.60), with $\dot{\tilde{\omega}}$ computed from

$$\dot{\tilde{\omega}} = \dot{\omega} + \dot{\Omega} = (\dot{\omega} + \dot{\Omega} \cos I) + \dot{\Omega} \sin I \left(\frac{\cos I - 1}{\sin I} \right) \tag{3.72}$$

with $\dot{\Omega} \sin I$ available from (3.43) and $\dot{\omega} + \dot{\Omega} \cos I$ from (3.47). The equations for $\dot{h}$ and $\dot{k}$ can then be computed to order one in I and e; the W term turns out to be $O(Ie)$ and is neglected:

$$na\dot{h} = -R \cos(\omega + f) + T \sin(\omega + f)$$
$$+ T \sin(\omega + u) + O(e^2) + O(Ie) \tag{3.73a}$$

and

$$na\dot{k} = R \sin(\omega + f) + T \cos(\omega + f)$$
$$+ T \cos(\omega + u) + O(e^2) + O(Ie). \tag{3.73b}$$

It turns out that (3.73) is identical to (3.61) without the out-of-plane terms with W; the analogue of (3.64) is then simply

$$na\dot{h} = -R \cos \lambda + 2T \sin \lambda - 3Th/2 - Rk$$
$$+ 3T(k \sin 2\lambda - h \cos 2\lambda)/2$$
$$- R(k \cos 2\lambda + h \sin 2\lambda) + O(e^2) + O(Ie) \tag{3.74a}$$

NON-SINGULAR ELEMENTS

and

$$na\dot{k} = R \sin \lambda + 2T \cos \lambda - 3Tk/2 + Rh$$
$$+ 3T(k \cos 2\lambda + h \sin 2\lambda)/2$$
$$+ R(k \sin 2\lambda - h \cos 2\lambda) + O(e^2) + O(Ie). \quad (3.74b)$$

As for the variation of the elements p, q which give the orientation of the osculating orbital plane, we have

$$\dot{p} = \frac{\dot{I} \sin \Omega}{\cos^2 I} - \tan I \, \dot{\Omega} \cos \Omega \quad (3.75a)$$

and

$$\dot{q} = \frac{\dot{I} \cos \Omega}{\cos^2 I} - \tan I \, \dot{\Omega} \sin \Omega. \quad (3.75b)$$

Using formulae (3.40) and (3.43), and neglecting the terms of order two or more in I we get

$$\dot{p} = \frac{W}{na\beta} (1 - e \cos u) \sin (\tilde{\omega} + f) + O(I^2) \quad (3.76a)$$

and

$$\dot{q} = \frac{W}{na\beta} (1 - e \cos u) \cos (\tilde{\omega} + f) + O(I^2) \quad (3.76b)$$

where only the W component appears, as expected.

By further expansion, with (3.62), in powers of e and neglecting the quadratic terms:

$$\dot{p} = \frac{W}{na} \left(\sin \lambda + \frac{k}{2} \sin 2\lambda - \frac{h}{2} \cos 2\lambda - \frac{3}{2}h \right) + O(e^2) + O(eI)$$
$$(3.77a)$$

and

$$\dot{q} = \frac{W}{na} \left(\cos \lambda + \frac{k}{2} \cos 2\lambda + \frac{h}{2} \sin 2\lambda - \frac{3}{2}k \right) + O(e^2) + O(eI).$$
$$(3.77b)$$

To find the equation for λ, we redefine ε as

$$\varepsilon + \int_{\bar{t}}^{t} n(t') \, dt' = M + \tilde{\omega} = M + \omega + \Omega = \lambda \qquad \dot{\varepsilon} = \dot{\eta} + \dot{\omega} + \dot{\Omega}$$
$$(3.78)$$

and the formula for $\dot{\varepsilon}$ will be very similar to (3.67), with only the W term changed because $\dot{\Omega}$ appears in (3.78):

$$\dot{\varepsilon} = \frac{R}{na}\left(-2 + \frac{3}{2}e\cos f\right) + \frac{T}{na}\frac{e}{2}\sin f$$

$$+ \frac{Wr}{na^2}\sin(\omega + f)\frac{1 - \cos I}{\sin I} + O(e^2) \qquad (3.79)$$

where $(1 - \cos I)/\sin I$ can be replaced by $I/2$ in the linear approximation.

3.4 SECULAR PERTURBATIONS

Equations for the variation of the elements, such as (3.30) or (3.32), are valid for every perturbing acceleration F, whatever its cause or size. However, non-gravitational perturbations belong to the class of 'small' perturbations, that is the ratio of F to the main monopole term of Earth's attraction is small, usually 10^{-6} or less (see table 2.1), so that the square of this ratio can be neglected.

When a perturbation is small, in the aforesaid sense, the differential equations for the osculating orbital elements can be treated in the perturbative way. Let $X(t)$ be the vector of the six orbital elements as a function of time, and let the differential equations be of the form

$$\dot{X} = F_0(X) + F_1(X)\delta \qquad (3.80)$$

where $\dot{X} = F_0(X)$ is the equation for the unperturbed Keplerian motion (or anyway for the reference orbit, e.g. precessing because of the J_{20} component of the gravity field of the Earth) and the perturbing acceleration has been written as $F = F_1\delta$, δ being a small parameter. Therefore the solution of (3.80) can be expanded in power series of δ:

$$X(t) = X_0(t) + X_1(t)\delta + X_2(t)\delta^2 + \ldots . \qquad (3.81)$$

If we compute the time derivative of (3.81), substitute into (3.80), expand $F_0(X)$ around X_0 and equate the coefficients of the same powers of δ we get

$$\dot{X}_0 = F_0(X_0) \qquad (3.82)$$

and

$$\dot{X}_1 = F_1(X_0) + \frac{\partial F_0}{\partial X}\bigg|_{X_0} X_1 \qquad (3.83)$$

for the zeroth-order and first-order equations. Equation (3.82) is just the equation for unperturbed motion; for Keplerian orbits it is $\dot{a} = \dot{e} = \dot{I} = \dot{\omega} = \dot{\Omega} = 0$, $\dot{M} = \mu^{1/2}a^{-3/2}$. Since we are interested only in small perturbations which give negligible second-order effects, the first-order perturbation equation (3.83) is the only one we need to solve. If the unperturbed solution $X_0(t)$ is a Keplerian orbit, the term $F_1(X_0)$ appearing in equation (3.83) simply means that in the equations for the variations of the Keplerian elements we have to substitute the elements of the unperturbed orbit osculating to the initial conditions. As an example, the equation for the variation of the semi-major axis in the simple case of an initially circular orbit of radius a_0 is easily obtained from (3.30):

$$\dot{a} = 2T/n_0 + O(\delta^2) \qquad (3.84)$$

where $n_0 = \mu^{1/2}a_0^{-3/2}$ is the mean motion of the initial osculating orbit and the transverse component T of the perturbing acceleration is also computed along the initial circular orbit with radius a_0. Equation (3.84) is correct to order two in the small parameter δ; this is because, for a Keplerian orbit, the vector $\dot{X}_0 = F_0(X_0)$ has only one non-zero component (i.e. the time derivative of the mean anomaly, which is a function of the semi-major axis only) and therefore the matrix $\partial F_0/\partial X|_{X_0}$ has only one non-zero element and the vector $\partial F_0/\partial X|_{X_0} \cdot X_1$ has only one non-zero component, again the one referring to the mean anomaly. This can be understood in the following way: the time derivative of the mean anomaly at any time is affected not only by the perturbations acting at that time but also by those which have been acting since the time corresponding to the initial conditions and have changed the semi-major axis (hence the mean motion). For this reason it is very convenient to write $M(t)$ as the sum of an angle ρ, whose time derivative is exactly the current mean motion n, plus an angle η such that:

$$M(t) = \rho + \eta \qquad (\dot{\rho} = n) \qquad (3.85)$$

(see (3.48)). In this way $\ddot{\rho} = \dot{n}$ (see (3.54)) takes care of the accumulated effect of the perturbations on the mean motion.

Similarly, it is convenient to write the mean argument of latitude $\lambda(t) = M + \omega$ (see (3.59)) as the sum of an angle λ_i whose time derivative is n, plus an angle ε such that

$$\lambda(t) = \lambda_i + \varepsilon \qquad (\dot\lambda_i = n). \tag{3.86}$$

The textbooks on celestial mechanics usually refer to the fact that for both ρ and λ_i a second-order differential equation has to be solved as the *double-integration problem*.

The first-order perturbation equation (3.83) is used in the following chapters of this book (with no mention of the omitted δ^2 terms) to compute both short-term and long-term effects.

When the perturbations are very small or well modelled, the short-periodic effects may not be important and we would like to be able to compute the long-term effects without being involved in the full computation. This is done with the *averaging method*. Of the osculating Keplerian elements, only M (or λ) is a rapid variable, that is it changes even for $\delta = 0$; the derivatives of all the others are $O(\delta)$. Therefore we average over one cycle of the rapid variable λ; let J be the vector of the other five osculating elements. The equations (3.80) are of the form

$$\dot\lambda = n(J) + K(J, \lambda) \tag{3.87a}$$

and

$$\dot J = L(J, \lambda) \tag{3.87b}$$

where K, L are functions 2π-periodic in λ giving the perturbations on λ and J (the other elements) respectively. The averaged equations are of the form

$$\dot J_A = L_A(J_A) \tag{3.88}$$

where L_A is the average of L over the interval from 0 to 2π:

$$L_A(J) = \frac{1}{2\pi} \int_0^{2\pi} L(J, \lambda)\, d\lambda. \tag{3.89}$$

The problem is to assess whether the solution $J_A(t)$ of the averaged equations (3.88) approximates the exact solution $J(t)$ of (3.87). If the variable λ is circulating, in the solution of (3.87), in a fast enough and regular enough way, then the average over λ and the running average over an interval of time comparable to the period of λ are close to one another. However, this is only a heuristic argument: the equations for J_A and J differ by terms of the order of δ, and after a time span of the order of $1/\delta$ orbital periods the solutions could be entirely different. The only rigorous theorem that can be applied in this case is the following.

Theorem: Assume that (3.87) is defined in a domain B, where the unperturbed frequency n does not vanish, and assume that δ is small enough ($\delta < \delta_0$). Then the distance between the solution $J_A(t)$ of the averaged equation (3.88) and the solution $J(t)$ of (3.87), with $J_A(0) = J(0)$, remains of the order of δ for a time interval of the order of $1/\delta$ orbital periods, i.e. there is a constant C such that

$$|J(t) - J_A(t)| < C\delta \qquad (3.90)$$

for $0 < t < P/\delta$ where P is the period $2\pi/n(J(0))$.

The proof can be found in Arnold (1983, Ch. 4). As in most theorems proved within the formalism of perturbation theory, it contains some unsatisfactory aspects. For example it refers to a 'small enough' δ, that is not quantitatively defined (nor easy to define). However, this result is very useful in practice: the most important example being the case in which the averaged equation is trivial, e.g.

$$\dot{J}_A = \frac{\delta}{2\pi} \int_0^{2\pi} L(J_A, \lambda) \, d\lambda = 0 \qquad (3.91)$$

In this case there is no 'secular' change in the elements apart from λ, and the theorem implies that J will remain constant, apart from oscillations with amplitude of the order of δ, for an interval of time of the order of $1/\delta$ orbital periods, which is a 'long' time.

This averaging method is widely employed in the following chapters, and it is indeed reliable enough for our purposes. However, the reader must be aware that there is a snag in the mathematical argument used to apply the abstract theorem to concrete cases. The theorem stated above applies only to equations of the form (3.87); equations (3.80) are in this form only provided that

(*a*) $\dot{X} = F_0$ is the equation for the unperturbed Keplerian orbit (in the case in which the effects of the Earth's oblateness are included in F_0, the other elements will change as well, e.g. the node precesses); and

(*b*) the perturbation $F_1(X)$ is a function of the the elements X only, and does not change with time.

The latter hypothesis unfortunately does not hold for many cases of interest in satellite geodesy: e.g. the perturbations due to the Earth's tesseral harmonics change with the Earth's rotation, and some non-gravitational perturbations change with the Sun's longitude (e.g. radiation pressure, drag) or even with both of them (e.g. the Earth-reflected radiation pressure).

Formally, if the perturbation F_1 depends upon time through periodic angular variables such as the Earth's rotation angle or the mean anomaly of the Sun, one can apply the averaging method with many frequencies, e.g. by averaging not only over M from 0 to 2π, but also over the mean anomaly of the Sun from 0 to 2π, etc. That is, the derivative of every osculating element will be described as a multiple Fourier series in the above-mentioned angles; then the constant term in this series can be computed, and it is assumed that the long-term behaviour is described by this 'secular' effect. This procedure is explicitly carried out in many cases in Chapter 4. Unfortunately, there is no rigorous estimate such as (3.90) of the difference between the 'secular' behaviour and the exact solution of the perturbed equations of motion.

The reasons why the method of averaging with more than one frequency is not rigorous, and in some cases may indeed fail to give a good estimate of the perturbative effects, have to do with some of the most fundamental problems of today's celestial mechanics. The reader can refer to Arnold (1983, Ch. 4) for a recent discussion and will find therein, as well as in most textbooks, that the reason for this failure of the averaged equations to describe correctly the long-term behaviour has to do with resonances. Whenever there are at least two 'fast' angular variables, say M_1 and M_2, with unperturbed frequencies n_1 and n_2, the ratio n_2/n_1 can be approximated by a rational number p/q. This amounts to saying that the real orbit will not spend the same amount of time in every surface element $dM_1\,dM_2$ of the M_1, M_2 surface, but will remain for a very long time close to a line in the same surface with equations

$$M_1 = M_1(0) + n_1 t$$
$$M_2 = M_2(0) + n_1(p/q)t. \quad (3.92)$$

Therefore running time averages along the true solution and space averages over M_1, M_2 ranging from 0 to 2π are very different, and the averaging fails to describe the long-term behaviour.

SECULAR PERTURBATIONS

However, in our case (non-gravitational perturbations on the orbit of artificial satellites) the only 'fast variables' are in many cases the satellite mean anomaly (or longitude) and the Sun's mean anomaly (or longitude); the ratio between the two mean motions is about 365 (for synchronous satellites) or more, and of the integers q and p used to approximate this ratio one at least must be quite large. As a result, the resonance is of a very high 'order'; that is in practice its effect will be very small, and a formula such as (3.90) does hold even if a rigorous mathematical proof is not available.

A little more troublesome is the case in which one of the frequencies involved in the perturbations is related to the rotation frequency of the Earth. In this case a resonance with small p, q can occur. This happens for the Earth's tesseral harmonics (see Kaula (1966) for a discussion); it can also happen for some of the perturbations discussed in this book, such as the Earth-reflected radiation pressure. However, these effects are very small, and unless a very strong resonance between the orbital period and the Earth's rotation period is present the effects will be described well enough by the averaging method. A more difficult case is the one of a 12-hour satellite in a very eccentric orbit (the so-called Molniya-type orbits), where strong albedo perturbations are felt near perigee. On the contrary, geostationary satellites are too far from the Earth to feel large perturbations from these causes (see table 2.1).

We conclude that the averaging method gives very reliable estimates of the long-term effects, provided some care is used in avoiding its use in resonant cases (for an example of non-gravitational perturbations with resonance, see Hori (1966)).

4

SOLAR RADIATION PRESSURE: DIRECT EFFECTS

4.1 SATELLITE–SOLAR RADIATION INTERACTION

By direct effect of the solar radiation on the satellite we mean the net acceleration resulting from the interaction (i.e. absorption, reflection, or diffusion) of the sunlight with each elementary surface of the spacecraft. Each photon carries an amount of momentum given by its energy (proportional to the light frequency) divided by the velocity of light, and this momentum can be exchanged during interaction with a solid surface. In this sense, the light behaves like a medium of material particles continuously emitted by the Sun, and indeed we shall show in Chapter 6 that several results obtained for radiation pressure are also valid for the drag force caused by a gaseous medium.

For each elementary surface dS of the spacecraft we call α, ρ and δ the fractions of incident sunlight which are absorbed, reflected and diffused respectively. They are also referred to as absorption, reflection and diffusion coefficients of the elementary surface dS (depending on its microscopic physical and chemical properties), and are related through the equation

$$\alpha + \rho + \delta = 1. \tag{4.1}$$

In the present context we shall make four simplifying assumptions:

(*a*) the absorbed light is not re-emitted (in fact the correspond-

SATELLITE–SOLAR RADIATION INTERACTION

ing energy is usually re-emitted as 'thermal' radiation of longer wavelength — see Chapter 5 for a discussion of the related perturbative effects);

(*b*) for a given direction, the intensity of diffused light is proportional to the cosine of the angle with the unit vector *n* perpendicular to dS (Lambert's law, i.e. spherical 'diffusion lobe');

(*c*) the reflection is perfectly specular;

(*d*) the elementary surface dS behaves like a linear combination of a black body, a perfect mirror and a Lambert diffuser (i.e. α, ρ and δ completely specify the optical properties of dS and for instance they do not depend on the angle β between *n* and the unit vector *S* from the surface to the Sun (see figure 4.1)).

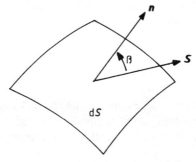

Figure 4.1

In these hypotheses the resulting elementary force dF caused on dS by the incident sunlight is given by

$$dF = -\frac{\Phi_\odot}{c}\left[(1-\rho)S + 2\left(\frac{\delta}{3} + \rho\cos\beta\right)n\right]dS|\cos\beta| \quad (4.2)$$

where $\Phi_\odot$ is the solar flux, c is the velocity of light and the absolute value of $\cos\beta = n \cdot S$ accounts for the cross section (in the direction perpendicular to the solar rays) of a surface that can be lightened on both sides. Expression (4.2) is obtained by adding together the elementary forces dF_α, dF_ρ and dF_δ caused on dS by the absorbed, reflected and diffused sunlight respectively. The elementary force dF_α is obviously directed along $-S$ and is proportional to the cross section, i.e.

$$dF_\alpha = -\frac{\Phi_\odot}{c}\alpha\, dS|\cos\beta|S. \quad (4.3)$$

The reflected light pushes the surface along $-n$, with a force proportional to $dS|\cos\beta|$ and also to the n component of the momentum transferred by the incident photons, i.e. to $2\cos\beta$:

$$dF_\rho = -\frac{\Phi_\odot}{c} 2\rho \cos\beta \, dS|\cos\beta|n. \qquad (4.4)$$

Finally, the force due to diffused light has components both along $-S$ and along $-n$; this because these photons can be assumed to be first absorbed, thus transferring to dS their linear momentum (which is directed as $-S$) and then re-emitted according to Lambert's law. The latter process produces a force that can be obtained by integrating the light intensity over the hemisphere external to dS. Since, at each angle θ with respect to n, the intensity is proportional to $\cos\theta$, this force is proportional to the following integral

$$\int_0^{\pi/2} \cos^2\theta (2\pi \sin\theta \, d\theta) = \tfrac{2}{3}\pi \qquad (4.5)$$

(where another factor $\cos\theta$ comes from the fact that for symmetry reasons only the n component of the diffused light produces a non-zero force when averaged over the hemisphere), divided by the integral

$$\int_0^{\pi/2} \cos\theta (2\pi \sin\theta \, d\theta) = \pi \qquad (4.6)$$

which is proportional to the total amount of diffused light. Hence, dF_δ turns out to be

$$dF_\delta = -\frac{\Phi_\odot}{c}\delta \, dS|\cos\beta|S - \frac{\Phi_\odot}{c}\tfrac{2}{3}\delta \, dS|\cos\beta|n \qquad (4.7)$$

and the expression (4.2) for dF is obtained as $dF = dF_\alpha + dF_\rho + dF_\delta$ (using also equation (4.1)).

If we want to study the total perturbative effect of solar radiation on the satellite's motion we can follow two basically different approaches. The most immediate one is purely numerical and can be divided in the following sequence of steps: first the spacecraft surface is decomposed into a finite number of small 'elementary' surfaces whose optical properties (specified by two of the three coefficients α, ρ and δ) are assumed to be known; then at any time and for each surface element the 'elementary' force given by expression (4.2) is computed for a specified relative position of the satellite with respect to the Sun; then the 'elementary' forces are summed up to give the total solar radiation perturbation vector acting on the satellite; finally, the latter is included in the right-hand side of the equations of motion that are eventually numerically integrated in order to obtain the satellite's position and velocity as functions of time. Apart from a possible inadequacy of the basic assumptions implicit in the model, the results will in any case be affected by the uncertainty in the knowledge of the spacecraft's attitude and of the optical properties of the surface. We will come back to this point in §4.4.

The second approach consists in using techniques typical of classical celestial mechanics (see Chapter 3) to decompose the effects of solar radiation pressure into long- and short-periodic terms. When one's aim is to analyse 'long' orbital arcs (i.e. a significant number of revolutions), the long-periodic terms can mix up with the purely gravitational effects and must be modelled as well as possible in order to extract all possible relevant information from the tracking data (e.g. for geophysical studies). On the contrary, the short-periodic effects can be thought of as a noise, superimposed onto the main signal, that one should try to filter out. Of course, in a second stage some model of the solar radiation perturbation must be included in the equations of motion, which can seldom be solved analytically: but in this way we know in advance where the most critical (e.g. long-periodic) orbital effects come from and what kind of phenomena must be accounted for in the model and/or carefully monitored when estimating the attainable accuracies. Thus *ad hoc* computer programs can be implemented to solve each particular problem in such a way as to minimise the errors introduced by the unavoidable incompleteness of every dynamical model. We are going to use the second approach in the next sections where both the method and the advantages will become apparent.

4.2 LONG-TERM EFFECTS ON SEMI-MAJOR AXIS

In order to attempt a theoretical investigation of the direct effects of solar radiation pressure on the motion of an artificial satellite we must first write down a general expression for the total solar radiation force depending on a few basic physical parameters. First of all, since the Earth–satellite distance is much smaller than the Earth–Sun distance, we expect the solar flux $\Phi_\odot$ on the satellite to be almost the same as that on the Earth ($\simeq 1.38 \times 10^6$ erg cm^{-2} s^{-1}). Hence the impinging photon flux will change mainly because of the eccentricity of the Earth's orbit around the Sun ($\simeq 0.017$), i.e. the solar radiation force on the spacecraft will contain a factor $(a_\odot^2/r_\odot^2)$, where $a_\odot = 1$ AU is the semi-major axis of the Earth's orbit and $r_\odot$ is its instantaneous distance from the Sun. Let us now first consider the case in which the spacecraft is axially symmetric with its symmetry axis fixed in the inertial space along the unit vector w (this is a suitable model even for a satellite of arbitrary shape provided it is rapidly spinning about a constant axis); whatever the direction S of the Sun, the resulting solar radiation force will lie in the plane generated by the vectors S and w. The force due to the absorbed sunlight will push the spacecraft along $-S$; the reflected and diffused light components will give rise to forces with both w and S components and the total acceleration on the spacecraft will be written as

$$F = -\frac{a_\odot^2}{r_\odot^2} [A(\psi)S + B(\psi)w] \quad (4.8)$$

where $\cos\psi = S \cdot w$.

It is worth stressing that if the spacecraft is not axially symmetric but is rapidly spinning (here 'rapidly' means that the spin period is much less than the orbital period, say at least 10 times shorter), all the previous discussion can also be applied by averaging the solar radiation force over one spin period of the spacecraft. This requires that the spin axis is fixed in the inertial space, which is true if it also coincides with the axis of maximum momentum of inertia and external torques are negligible—otherwise, the spacecraft can precess, and expression (4.8) for the solar radiation force still applies only if the precession period is much shorter than the orbital one. Of course in this case the w unit vector is the axis around which the spin axis is precessing. In the case of an axially symmetric spacecraft rotating around an axis different from the axis of maximum momentum of inertia, the ratio between the spin period and the period of free precession is of the order of $(C-A)/C$, C and A being the two principal moments of inertia (with $C > A$). A

LONG-TERM EFFECTS ON SEMI-MAJOR AXIS

spacecraft with very different principal moments of inertia does not necessarily precess in a regular way; it might also tumble in a very chaotic way and in this case there is no hope of modelling the effects of the solar radiation force with any reasonable accuracy unless the attitude is continuously monitored.

Let us first investigate the effects of the solar radiation force as given by equation (4.8) on the semi-major axis of the satellite. As shown in §3.2, equation (3.30), only the R and T components of the perturbing force (i.e. the radial and in-plane transverse component respectively) contribute to the time variation of the semi-major axis

$$\dot{a} = \frac{2e \sin f}{n(1 - e^2)^{1/2}} R + \frac{2(1 + e \cos f)}{n(1 - e^2)^{1/2}} T. \qquad (4.9)$$

In an orthogonal reference frame with the origin in the centre of the Earth, the x-axis pointing towards the perigee of the satellite orbit and the z-axis directed as the satellite orbital angular momentum, the radial, in-plane transverse and out-of-plane unit vectors e_R, e_T, e_W are written as

$$e_R = (\cos f, \sin f, 0)$$
$$e_T = (-\sin f, \cos f, 0) \qquad (4.10)$$
$$e_W = (0, 0, 1).$$

We have assumed that the satellite symmetry (or rotation) axis is fixed in the inertial space. This symmetry (or rapid rotation) hypothesis, plus the two assumptions that the optical properties of the spacecraft surface are constant in time and that the spacecraft temperature may not be constant but may change only as function of the Sun's position, imply that the solar radiation force F expressed by equation (4.8) depends only on S (the position of the Sun), i.e.

$$F = F(S) \qquad (4.11)$$

with components $F_x(S)$, $F_y(S)$, $F_z(S)$ in the frame of reference defined above. Since $R = F(S) \cdot e_R$ and $T = F(S) \cdot e_T$, equation (4.9) becomes

$$\dot{a} = -\frac{2 \sin f}{n(1 - e^2)^{1/2}} F_x(S) + \frac{2(e + \cos f)}{n(1 - e^2)^{1/2}} F_y(S). \qquad (4.12)$$

An easy way of separating short-periodic effects ('short' means comparable with the satellite orbital period) from long-periodic (~ 1 year) or secular effects, consists in expressing $\dot{a}$ as a Fourier series having as arguments the mean anomalies of the satellite and the Sun, M and M_1 respectively

$$\dot{a} = \sum_{k_1,k_2 \in Z} \dot{a}_{k_1 k_2} \exp[i(k_1 M_1 + k_2 M)] \tag{4.13}$$

where the coefficients $\dot{a}_{k_1 k_2}$ are given by

$$\dot{a}_{k_1 k_2} = \frac{1}{(2\pi)^2} \int_0^{2\pi} \int_0^{2\pi} \dot{a} \exp[-i(k_1 M_1 + k_2 M)] \, dM_1 \, dM. \tag{4.14}$$

The long-term coefficients do not contain M, i.e. they have $k_2 = 0$:

$$\dot{a}_{k_1 0} = \frac{1}{(2\pi)^2} \int_0^{2\pi} \int_0^{2\pi} \dot{a} \exp(-i k_1 M_1) \, dM_1 \, dM. \tag{4.15}$$

Using equation (4.12) for $\dot{a}$ they become

$$\dot{a}_{k_1 0} = \frac{1}{(2\pi)^2} \frac{2}{n(1-e^2)^{1/2}}$$

$$\times \left(-e \int_0^{2\pi} F_x(S) \exp(-i k_1 M_1) \, dM_1 \int_0^{2\pi} \sin f \, dM \right.$$

$$\left. + \int_0^{2\pi} F_y(S) \exp(-i k_1 M_1) \, dM_1 \int_0^{2\pi} (e + \cos f) \, dM \right). \tag{4.16}$$

The sine and cosine of the true anomaly can be expressed as functions of the mean anomaly as in equation (3.62), where terms of higher order in e are also of the kind $\sin kM$, $\cos kM$ (with k a positive integer) in the expansion of $\sin f$ and $\cos f$ respectively. Making use of these expansions we have

$$\int_0^{2\pi} \sin f \, dM = 0 \tag{4.17}$$

and

$$\int_0^{2\pi} (e + \cos f) \, dM = 0 \tag{4.18}$$

and therefore

$$\dot{a}_{k_1 0} = 0. \tag{4.19}$$

We can conclude with the following

Theorem: If the total solar radiation force acting on the spacecraft can be expressed in the general form (4.11), no long-term effect in semi-major axis will appear to any order in the orbital eccentricity.

It is worth noting that in proving the theorem we have used a two-body approximation, assuming a point-mass Earth and a spacecraft perturbed by solar radiation only. In this approximation the reference directions of the satellite angular momentum and of the direction of perigee are fixed. If we now take into account the J_{20} term, namely the oblateness of the Earth, it will affect both the perigee and the node of the satellite orbit, so that a solar radiation perturbation producing long-term effects will appear. However, this perturbation will be of the order of the solar radiation force times J_{20}, i.e. about 10^3 times smaller than the solar radiation force itself. Since the accuracy in the radiation pressure model is not all that good (see §4.4), these effects can usually be neglected.

In the case of passive satellites of spherical shape like Starlette or LAGEOS, one expects only short-period effects in semi-major axis to appear, apart from the above-mentioned effects of the first order in the J_{20} coefficient. Moreover, they are quite easy to model because of the spherical shape of the spacecraft. However, we are also interested in accurate orbit determination for satellites which are not passive 'cannon balls'. Whenever a satellite is active, its shape is usually complex, and often a quite essential part of the spacecraft is an antenna, allowing it to communicate with ground stations. In most cases the antenna is a directional one, allowing radiowaves to be beamed towards the Earth. With such an antenna, the hypotheses we have used to prove the previous theorem are no longer applicable. Let b be the direction of the radiowave beam from the spacecraft to some point on the Earth; then b has to change with the satellite anomaly, and directional pointing is possible only if the antenna mount rotates around an axis w_A that is more or less perpendicular to the orbital plane. Then the radiation pressure acceleration cannot be of the form (4.11). If N_A is the unit vector along an axis of symmetry of the antenna, the acceleration F_A due to the Sun's radiation on the antenna will be

$$F_A = \mathscr{G}(f)S + \mathscr{F}(f)N_A \qquad (4.20)$$

where $\mathscr{G}$, $\mathscr{F}$ are functions of the true anomaly of the satellite (and of the Sun's position as well). To compute the latter we will make the following simplifying assumptions:

(a) The spacecraft body feels a radiation pressure acceleration given by expression (4.11); the mutual shadowing between the body and the antenna is neglected and further complications such as multiple reflections are also neglected.

(b) The antenna spin axis w_A is close to the perpendicular to the orbit plane e_W; the small angle between the two will be called η.

(c) The radiowaves are beamed in the direction b that is close to the direction of the Earth's centre $(-e_R)$; the small misalignment angle in the orbital plane is called ξ. ξ cannot be assumed zero, not only because of possible pointing errors, but also because the beam can be accurately pointed towards a ground station which is not directly below the satellite. We also assume that $N_A = b$ (this is not true for every possible antenna design, but is of course the simplest case).

(d) The satellite orbit has a small eccentricity e.

In these hypotheses we can expand the expression (4.20) into a power series in η, ξ and e, and neglect the terms of second (or higher) degree. We first expand equation (4.9) for $\dot{a}$ in powers of e, by using equations (3.62):

$$\dot{a} = \frac{2}{n} eR_A \sin M + \frac{2}{n} (1 + e \cos M) T_A \qquad (4.21)$$

where

$$R_A = F_A \cdot e_R \qquad T_A = F_A \cdot e_T. \qquad (4.22)$$

Then we go back to expression (4.20) and write S and $N_A = b$ in the e_R, e_T, e_W moving frame. We have

$$S = \begin{pmatrix} B_1 \cos f + B_2 \sin f \\ B_2 \cos f - B_1 \sin f \\ B_3 \end{pmatrix} \qquad (4.23)$$

where B_1, B_2 and B_3 are functions of time only through the mean longitude of the Sun; and

$$b = \begin{pmatrix} -1 \\ -\xi \\ \eta \cos(\zeta - f) \end{pmatrix} \qquad (4.24)$$

where ζ is the angle between the projection of w_A on the orbital plane and the direction of the perigee (because the latter is the origin from which the true anomaly is computed; see figure 4.2). We can now write equation (4.21) for the variation of the semi-major axis as follows:

$$\frac{n}{2}\dot{a} = \mathcal{G}(M)(B_2 \cos M - B_1 \sin M) - \xi\mathcal{F}(M)$$
$$+ eB_2 \mathcal{G}(M) - e \sin M \, \mathcal{F}(M)$$
$$+ e\mathcal{G}(M)[B_2(\cos 2M - 1) - B_1 \sin 2M]$$
$$+ e\mathcal{G}_1(M)(B_2 \cos M - B_1 \sin M) \qquad (4.25)$$

where terms of the order of e^2, $e\xi$, ξ^2, have been neglected and $\mathcal{G}_1(M)$ is defined by the following expansion

$$\mathcal{G}(f) = \mathcal{G}(M) + e\mathcal{G}_1(M) + O(e^2). \qquad (4.26)$$

For $\sin f$ and $\cos f$, D'Alembert expansions (3.62) in $\sin M$ and $\cos M$ have been used.

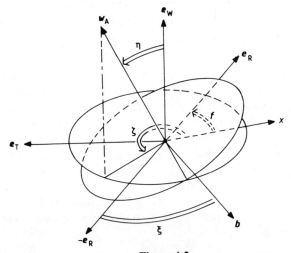

Figure 4.2

As we can see in equation (4.25) all but one of the terms contributing to the variation of the semi-major axis because of the solar radiation force on the antenna are demultiplied by a small parameter (e or ξ). The $\mathcal{G}$ term, which appears at zero order in the small parameters, is a long-period one since it contains the longitude of the Sun. Let us now evaluate it in the simple case of a flat antenna; in this case the $\mathscr{G}$ function can be written as

$$\mathscr{G}(f) = R' \cos \beta + R'' |\cos \beta| \qquad (4.27)$$

(where now $\cos \beta = N_A \cdot S$). The two constants R' and R'' take into account the possibility of different optical properties and/or temperatures between the two sides of the antenna. Since we are interested in the long-period effects of the term of zero order in the small parameters we compute $\cos \beta$ for $\eta = \xi = e = 0$

$$\cos \beta = N_A \cdot S = -B_1 \cos M - B_2 \sin M \qquad (4.28)$$

and average over the mean anomaly of the satellite. The resulting integral turns out to be zero

$$\int_0^{2\pi} \mathscr{G}(M)(B_2 \cos M - B_1 \sin M) \, dM$$

$$= R'L^2 \int_\tau^{\tau - 2\pi} \cos(\tau - M) \sin(\tau - M) \, d(\tau - M)$$

$$+ R''L^2 \int_\tau^{\tau - 2\pi} |\cos(\tau - M)| \sin(\tau - M) \, d(\tau - M) = 0 \qquad (4.29)$$

where we have introduced the new variables L and τ defined by $B_1 = L \cos \tau$ and $B_2 = L \sin \tau$, and both the resulting integrals vanish because they are integrals of odd functions over one period of the integration variable. Therefore, all the long-period effects on the semi-major axis appearing because of the solar radiation force on a flat antenna are demultiplied by the small parameters e and ξ (i.e. the orbital eccentricity and the misalignment angle). For satellites having small eccentricities, like geostationary satellites, the misalignment of the antenna can be the major factor responsible for long-period effects in semi-major axis, whose order of magnitude is

$$\dot{a}_{lp} \simeq \frac{2}{n} \left(\frac{\mathscr{A}}{\mathscr{M}} \frac{\Phi_\odot}{c} \right) \times \frac{\text{antenna cross section}}{\text{satellite cross section}} \times \text{misalignment angle} \qquad (4.30)$$

where $\mathscr{A}/\mathscr{M}$ is the area-to-mass ratio of the satellite and the product $(\mathscr{A}\Phi_\odot/\mathscr{M}c)$ gives the order of magnitude of the total solar radiation acceleration, which is demultiplied by the small parameter ξ (to be substituted by e, if the latter is larger).

LONG-TERM EFFECTS ON SEMI-MAJOR AXIS

If the despun antenna is not flat, but not too complicated in shape, in such a way that $\mathscr{G}(M)$ can still be expressed by a formula similar to expression (4.27), containing only the cosine of the angle between the direction of the Sun and the symmetry axis of the antenna, the previous result is still true—and if the eccentricity is small efforts should be made to try to keep the misalignment angle as small as possible. In §4.4 we shall discuss the case of some satellites of interest for geophysical studies.

We shall now compute the effects of the solar radiation perturbation on the mean argument of latitude, $\lambda = M + \omega$. It can be written (as in equation (3.65)) as the sum of two terms

$$\lambda = \lambda_i + \varepsilon \tag{4.31}$$

where λ_i is the value of λ for an ideal satellite moving with the osculating mean motion at any time

$$\lambda_i(t) = \int_0^t n(t') \, dt' \tag{4.32}$$

and $\varepsilon = \lambda - \lambda_i$ gives the difference between the mean longitude of the actual satellite and its value for the ideal one. The osculating mean motion and semi-major axis satisfy, at any time, Kepler's third law

$$n^2 a^3 = \text{constant}. \tag{4.33}$$

Differentiating with respect to time we obtain

$$\dot{n}(t) = -\frac{3}{2} \frac{n(t)}{a(t)} \dot{a}(t). \tag{4.34}$$

Thus, at the first perturbative order, $\dot{n}(t)$ and $\dot{a}(t)$ (say, due to solar radiation) are related through the equation

$$\dot{n}(t) = -\frac{3}{2} \frac{n(0)}{a(0)} \dot{a}(t) \tag{4.35}$$

where $n(0)$ and $a(0)$ are the initial values of the mean motion and the semi-major axis respectively. From the definition of λ_i and equation (4.35), we have the following differential equation

$$\ddot{\lambda}_i(t) = -\frac{3}{2} \frac{n(0)}{a(0)} \dot{a}(t) \tag{4.36}$$

which immediately gives the (second) derivative of λ_i once $\dot{a}(t)$ has been computed.

As an example, for a spacecraft body satisfying the hypotheses in which equation (4.19) (i.e. no long-periodic effect in semi-major axis) has been proven to be true, we can say that no long-periodic effects arise in λ_i either. However, the effects in longitude caused by the ε term (see equation (4.31)), i.e. not coming from the variation of semi-major axis, have still to be computed; this will be done in the next section.

In the case of a spinning antenna with a one-to-one resonance between the spin and the orbital period, for which we have proved equation (4.29) and concluded that long-periodic effects in semi-major axis are demultiplied by a (usually) small parameter, the effects in λ_i are easily obtained as well. In particular, for satellites with very small orbital eccentricity, our previous estimate of the long-term perturbing force due to the antenna also yields an order-of-magnitude estimate of the long-period effects in λ_i:

$$(a\ddot{\lambda}_i)_{lp} \simeq 3\, \frac{\mathcal{A}\Phi_\odot}{\mathcal{M}c} \times \frac{\text{antenna cross section}}{\text{satellite cross section}} \times \text{misalignment angle} \quad (4.37)$$

(through equations (4.30) and (4.36)). Apart from the misalignment angle involved in the particular problem that we are investigating, it is worth noting that the along-track longitude acceleration is always three times greater than the perturbing force per unit mass. This 'paradoxical' result comes from the combination of equation (4.36) with the Gauss equation (4.21) for the variation of semi-major axis.

Although formulae like (4.30) and (4.37) give only an order of magnitude, they are very useful because they can give an idea of the long-term effects of solar radiation for particular satellites equipped with antennae, and therefore allow us to easily estimate the accuracy that can be achieved in recovering the parameters of geophysical interest. We shall see some examples in §4.4.

4.3 LONG-TERM EFFECTS ON THE OTHER ORBITAL ELEMENTS

The equation for the variation of ε (defined by equation (4.31)), also valid for zero eccentricity, has been given in §3.3 (equation 3.67). It contains the R, T and W components of the perturbing force and the expansion in powers of the satellite eccentricity has been stopped at order one.

LONG-TERM EFFECTS ON OTHER ORBITAL ELEMENTS

Let us now consider a spacecraft body satisfying the three assumptions discussed in §4.2, i.e. let us assume that the solar radiation force can be written as a function of the Sun position vector only:

$$F = F(S). \qquad (4.38)$$

In the orthogonal reference frame with the x-axis pointing towards the perigee of the satellite orbit and the z-axis directed as the orbital angular momentum vector, the e_R, e_T and e_W unit vectors are given by expressions (4.10) and the R, T, and W components of the force (4.38) are

$$R = F(S) \cdot e_R = F_x(S)\cos f + F_y(S)\sin f$$
$$T = F(S) \cdot e_T = -F_x(S)\sin f + F_y(S)\cos f \qquad (4.39)$$
$$W = F(S) \cdot e_W = F_z(S).$$

We shall now prove that a force depending only on S does not give rise to any long-period effects in ε to zero order in the satellite eccentricity. Using D'Alembert expansions (3.62) for $\cos f$ and $\sin f$, equation (3.67) to zero order in e becomes

$$\dot{\varepsilon} = -\frac{2}{na} R - \frac{1}{na} \cot I \sin(\omega + M) W + O(e). \qquad (4.40)$$

In order to compute long-period effects we express $\dot{\varepsilon}$ in Fourier series of the mean anomalies of the Sun M_1 and of the satellite M

$$\dot{\varepsilon} = \sum_{k_1, k_2 \in Z} \dot{\varepsilon}_{k_1 k_2} \exp[i(k_1 M_1 + k_2 M)] \qquad (4.41)$$

as we did for $\dot{a}$ in §4.2. The Fourier coefficients $\dot{\varepsilon}_{k_1 k_2}$ are given by

$$\dot{\varepsilon}_{k_1 k_2} = \frac{1}{(2\pi)^2} \int_0^{2\pi} \int_0^{2\pi} \dot{\varepsilon} \exp[-i(k_1 M_1 + k_2 M)] \, dM_1 \, dM \qquad (4.42)$$

and long-period effects will have $k_2 = 0$ and coefficients $\dot{\varepsilon}_{k_1 0}$. Long-period effects are given by

$$\dot{\varepsilon}_{k_1 0} = \frac{1}{(2\pi)^2} \int_0^{2\pi} \int_0^{2\pi} \left(-\frac{2}{na}\right) [F_x(S)\cos f + F_y(S)\sin f]$$
$$\times \exp(-ik_1 M_1) \, dM_1 \, dM$$
$$+ \frac{1}{(2\pi)^2} \int_0^{2\pi} \int_0^{2\pi} \left(-\frac{1}{na}\right) \cot I \sin(\omega + M) F_z(S)$$
$$\times \exp(-ik_1 M_1) \, dM_1 \, dM. \qquad (4.43)$$

To zero order in the eccentricity cos f and sin f can be replaced by cos M and sin M respectively, and therefore we have

$$\dot{\varepsilon}_{k;0} = 0 \qquad (4.44)$$

because of the factors containing the integrals from 0 to 2π of sin M and cos M. We conclude that all the long-period effects in longitude coming from the ε term for a spacecraft body such that the solar radiation force can be written as in (4.38), are demultiplied by a factor e. Moreover, it is worth stressing that such effects do not accumulate quadratically since we need to integrate only once to get the longitude effect, while equation (4.36) must be integrated twice and therefore the λ_i perturbation grows quadratically with time. A useful formula to obtain an order-of-magnitude estimate of the long-periodic effect in ε is the following:

$$(a\dot{\varepsilon})_{\text{lp}} \simeq \frac{1}{2\pi} \left(\frac{\mathscr{A}\Phi_\odot}{\mathscr{M}c} \right) eP \qquad (4.45)$$

where P is the orbital period of the satellite and $\mathscr{A}\Phi_\odot/\mathscr{M}c$ gives the order of magnitude of the solar radiation force per unit mass of the satellite.

However, formula (4.45) fails whenever the satellite has an outside despun antenna. The solar radiation force on the antenna can be expressed as in (4.20), and the components of S and N_A in the moving e_R, e_T, e_W frame are given by formulae (4.23) and (4.24) respectively. The R_A and W_A components of the force (4.20) to zero order in both e and η are

$$R_A = F_A \cdot e_R = \mathscr{G}(M)(B_1 \cos M + B_2 \sin M) - \mathscr{F}(M) \qquad (4.46)$$

and

$$W_A = F_A \cdot e_W = \mathscr{G}(M)B_3 \qquad (4.47)$$

and the equation corresponding to equation (4.40) for the antenna is

$$\dot{\varepsilon}^A = -\frac{2}{na} [\mathscr{G}(M)(B_1 \cos M + B_2 \sin M) - \mathscr{F}(M)]$$

$$-\frac{1}{na} \cot I \sin(\omega + M) \mathscr{G}(M)B_3 + O(\eta) + O(e). \qquad (4.48)$$

LONG-TERM EFFECTS ON OTHER ORBITAL ELEMENTS

The corresponding long-period Fourier coefficients $\dot{\varepsilon}^A_{k_10}$ are

$$\dot{\varepsilon}^A_{k_10} = \frac{1}{(2\pi)^2} \int_0^{2\pi} \int_0^{2\pi} \left(-\frac{2}{na}\right) \exp(-ik_1 M_1)$$
$$\times [\mathcal{G}(M)(B_1 \cos M + B_2 \sin M) - \mathcal{F}(M)] \, dM_1 \, dM$$
$$+ \frac{1}{(2\pi)^2} \int_0^{2\pi} \int_0^{2\pi} \left(-\frac{1}{na} \cot I\right) \exp(-ik_1 M_1) B_3$$
$$\times \sin(\omega + M) \mathcal{G}(M) \, dM_1 \, dM \qquad (4.49)$$

and they do not average out because of the $\mathcal{G}(M)$, $\mathcal{F}(M)$ factors which come from the solar radiation force (4.20) on the antenna. In other words, the long-periodic effect on ε is not zero because the solar radiation force on the antenna contains the mean anomaly of the satellite. The order-of-magnitude of this effect in ε can be estimated to be

$$(a\dot{\varepsilon}^A)_{lp} \simeq \frac{1}{2\pi} \left(\frac{\mathcal{A}\Phi_\odot}{\mathcal{M}c}\right) P \times \frac{\text{antenna cross section}}{\text{satellite cross section}} \qquad (4.50)$$

with the same notation as in equation (4.45).

We can easily prove that in the case of a spacecraft body for which the perturbation due to solar radiation can be written in the form (4.38), so that $W = F_z(S)$, no long-periodic effects appear on the inclination and the node of the satellite to zero order in the eccentricity. To zero order in e, equations (3.40) and (3.43) for the variation of I and Ω become

$$\dot{I} = \frac{W}{H} a \cos(\omega + M) + O(e) \qquad (4.51a)$$

and

$$\dot{\Omega} = \frac{1}{\sin I} \frac{W}{H} a \sin(\omega + M) + O(e) \qquad (4.51b)$$

where H is the satellite orbital angular momentum. $\dot{I}$ and $\dot{\Omega}$ can be expanded in Fourier series of M_1 and M, as we have done for $\dot{a}$ (see expressions (4.13) and (4.14)). The long-periodic terms have $k_2 = 0$, so that their coefficients are $\dot{I}_{k_10}$, $\dot{\Omega}_{k_10}$. Using equations (4.51) with $W = F_z(S)$ and keeping only the zero order in e we get

$$\dot{I}_{k_10} = \frac{1}{(2\pi)^2} \int_0^{2\pi} \frac{a}{H} F_z(S) \exp(-ik_1 M_1) dM_1 \int_0^{2\pi} \cos(\omega + M) \, dM = 0$$
$$(4.52a)$$

and

$$\dot{\Omega}_{k_1 0} = \frac{1}{(2\pi)^2} \int_0^{2\pi} \frac{a}{H \sin I} F_z(S) \exp(-ik_1 M_1) \, dM_1$$

$$\times \int_0^{2\pi} \sin(\omega + M) \, dM = 0 \qquad (4.52b)$$

again because of the integrals of sin M and cos M. Therefore, the long-periodic effects on I and Ω are demultiplied by a factor e. An order-of-magnitude estimate of the long-periodic effect in inclination is given by the formula

$$(a\dot{I})_{\rm lp} \simeq \frac{1}{2\pi} \left(\frac{\mathcal{A}\Phi_\odot}{\mathcal{M}c} \right) eP. \qquad (4.53)$$

It is worth noting that $\mathcal{A}\Phi_\odot/\mathcal{M}c$ is the order of magnitude of the total acceleration due to solar radiation, whereas only its out-of-plane component contributes to the effect in inclination. In the same way in equation (4.30) $\mathcal{A}\Phi_\odot/\mathcal{M}c$ was an upper estimate, since only the in-plane components of the solar perturbation contribute to the variation of the semi-major axis.

If the satellite has also an outside rotating antenna on which the solar radiation perturbation can be written as in (4.20), long-periodic effects on the inclination and the node appear to zero order in the small parameters. Equations corresponding to equations (4.51) for the antenna are

$$\dot{I}^A = \frac{W_A}{H} a \cos(\omega + M) + O(e) \qquad (4.54a)$$

and

$$\dot{\Omega}^A = \frac{W_A}{H \sin I} a \sin(\omega + M) + O(e) \qquad (4.54b)$$

where $W_A = F_A \cdot e_W$ is the out-of-plane component of the perturbing acceleration on the antenna given by (4.47) to zero order in e and η. The long-period coefficients of the Fourier expansions for $\dot{I}^A$, $\dot{\Omega}^A$ are

$$\dot{I}^A_{k_1 0} = \frac{1}{(2\pi)^2} \int_0^{2\pi} \int_0^{2\pi} \frac{a}{H} \exp(-ik_1 M_1) B_3 \mathcal{G}(M) \cos(\omega + M) \, dM_1 \, dM$$

$$(4.55a)$$

and
$$\dot{\Omega}^A_{k,0} = \frac{1}{(2\pi)^2} \int_0^{2\pi} \int_0^{2\pi} \frac{a}{H \sin I}$$
$$\times \exp(-ik_1 M_1) B_3 \mathscr{G}(M) \sin(\omega + M) \, dM_1 \, dM. \quad (4.55b)$$

As we can see, they are no longer zero because of the $\mathscr{G}(M)$ term which appears in the force expression (4.20) and is a function of the satellite mean anomaly. An order-of-magnitude estimate of the effect in inclination due to the antenna can be given by the following formula

$$(a\dot{I}^A)_{lp} \simeq \frac{1}{2\pi} \left(\frac{\mathscr{A} \Phi_\odot}{\mathscr{M} c} \right) P \frac{\text{antenna cross section}}{\text{satellite cross section}}. \quad (4.56)$$

In §4.4 we shall discuss the limitations due to this effect on polar motion determinations when satellites equipped with antennae are used. We note that in the case of satellites with small inclination we must use non-singular elements for $I = 0$. Since we usually deal with satellites with small eccentricy as well, equations (3.77) can be used, since they are non-singular for both $e = 0$ and $I = 0$. Employing the usual techniques for the computation of the long-period effects we would get similar results, i.e. that the spacecraft body contributes to long-periodic effects in I and Ω only to first order in e, while there are long-periodic effects also to zero order in e caused by the solar radiation pressure on the antenna.

Finally, let us compute the effect of solar radiation perturbation on the satellite eccentricity. We shall use the elements non-singular for $e = 0$, namely $h = e \sin \omega$, $k = e \cos \omega$, and equations (3.64) for $\dot{h}$ and $\dot{k}$ where terms of second order in e are neglected. Firstly we have to compute the long-periodic effects of terms of zero order in e in the hypothesis of a perturbation of the form $\mathbf{F} = \mathbf{F}(S)$. As we shall see, they do not vanish like those for the other orbital elements. Equations (3.64) to zero order in e become

$$\dot{h} = \frac{1}{na} [-R \cos(\omega + M) + 2T \sin(\omega + M)] + O(e) \quad (4.57a)$$
and
$$\dot{k} = \frac{1}{na} [R \sin(\omega + M) + 2T \cos(\omega + M)] + O(e) \quad (4.57b)$$

where R and T are given by expressions (4.39). As usual $\dot{h}$ and $\dot{k}$ have to be expanded in Fourier series of M_1 and M, and the long-periodic terms have coefficients (to zero order in e)

$$\dot{h}_{k_10} = \frac{1}{(2\pi)^2} \int_0^{2\pi} \int_0^{2\pi} \left(-\frac{1}{na}\right) \exp(-ik_1 M_1)$$
$$\times (F_x(S)\cos M + F_y(S)\sin M)\cos(\omega + M) \, dM_1 \, dM$$
$$+ \frac{1}{(2\pi)^2} \int_0^{2\pi} \int_0^{2\pi} \left(\frac{2}{na}\right) \exp(-ik_1 M_1)$$
$$\times (F_y(S)\cos M - F_x(S)\sin M)\sin(\omega + M) \, dM_1 \, dM \quad (4.58)$$

and

$$\dot{k}_{k_10} = \frac{1}{(2\pi)^2} \int_0^{2\pi} \int_0^{2\pi} \left(\frac{1}{na}\right) \exp(-ik_1 M_1)$$
$$\times (F_x(S)\cos M + F_y(S)\sin M)\sin(\omega + M) \, dM_1 \, dM$$
$$+ \frac{1}{(2\pi)^2} \int_0^{2\pi} \int_0^{2\pi} \left(\frac{2}{na}\right) \exp(-ik_1 M_1)$$
$$\times (F_y(S)\cos M - F_x(S)\sin M)\cos(\omega + M) \, dM_1 \, dM. \quad (4.59)$$

These coefficients do not average out because of the integrals of $\cos^2 M$ and $\sin^2 M$ (instead of $\cos M$ and $\sin M$) which are not zero. Therefore we expect to have long-periodic (i.e. approximately yearly) effects in the eccentricity, coming from the solar radiation perturbation on the spacecraft body, even in the hypothesis that it can be written in the form (4.38) and also for a satellite in a circular orbit. A rough formula to estimate this long-periodic effect in eccentricity is the following:

$$(a\dot{e})_{lp} \simeq \frac{1}{2\pi} \left(\frac{\mathcal{A}\Phi_\odot}{\mathcal{M}c}\right) P. \quad (4.60)$$

Again, $\mathcal{A}\Phi_\odot/\mathcal{M}c$ is an upper limit for the perturbing force since only the in-plane components affect the satellite eccentricity. This effect shows up as an oscillation of the satellite distance over one orbital period; the amplitude of this oscillation will vary roughly with the rate (4.60) and a yearly periodicity. There are also zero-order long-period effects in eccentricity coming from the solar radiation force on the antenna, as can easily be shown with the techniques used so far. However, they are smaller than the effects estimated in formula (4.60), because of the usually smaller cross section of the antenna with respect to the spacecraft body.

4.4 CONCLUSIONS AND EXAMPLES

In this section we summarise the effects of direct solar radiation for some particular satellites of interest for geodetic and geophysical studies. We will consider geosynchronous satellites of various kinds, GPS satellites, SEASAT (or ERS-1) and LAGEOS.

Geostationary satellites (i.e. satellites with orbital period equal to the rotation period of the Earth, small eccentricity, and small inclination to the equatorial plane of the Earth) are resonant with the coefficients of the Earth's gravity field with low degree l and even values of the difference $(l - m)$; this causes the satellite's longitude to librate with a very long period (of the order of 10^3 days) about two stable equilibrium positions, implying that for orbital arcs shorter than several months the longitude perturbation caused by the resonant geopotential terms accumulates almost quadratically with time (see for example Kaula (1966, §3.6) and also §7.1). Therefore orbital data of geosynchronous satellites can be used to recover the resonant coefficients; the achievable accuracy depends on both the accuracy in modelling the solar radiation perturbation and the accuracy of the tracking data. We shall see that the main limitation often comes from the long-periodic effects of solar radiation rather than from the available tracking technologies.

In the case of three-axis stabilised geosynchronous satellites which continuously point towards the Earth, there is no rapidly spinning spacecraft body, and referring to the treatment of the previous sections the whole satellite can be viewed as a big antenna. For the effect in semi-major axis (hence in λ_i), we are not necessarily allowed to use for $\ddot{\lambda}_i$ a formula like (4.37), where the small parameter of the antenna misalignment appears. This is because, according to equation (4.25), the zero-order term can be proved to give no long-periodic effect provided the component of the radiation force on the antenna can be written as in expression (4.27), which of course is very unlikely for a spacecraft of complex shape. Therefore, a pessimistic but more reliable estimate can be obtained from the formula

$$a\ddot{\lambda}_i \simeq 3 \frac{\mathscr{A}\Phi_\odot}{\mathscr{M} c}. \tag{4.61}$$

Moreover, there is also an effect in longitude coming from the ε

term and of the order of

$$(a\Delta\varepsilon) \simeq \frac{1}{2\pi}\left(\frac{\mathscr{A}\Phi_\odot}{\mathscr{M}c}\right)P^2 \times \left(\frac{\Delta t}{1\text{ day}}\right) \qquad (4.62)$$

(see formula (4.50)), which, however, does not accumulate quadratically with time. The effect in inclination can be also estimated to be

$$(a\Delta I) \simeq \frac{1}{2\pi}\left(\frac{\mathscr{A}\Phi_\odot}{\mathscr{M}c}\right)P^2 \times \left(\frac{\Delta t}{1\text{ day}}\right) \qquad (4.63)$$

(see formula (4.53)), and the effect in eccentricity is given by equation (4.60). We give the corresponding figures for a three-axis stabilised satellite with an area-to-mass ratio of $0.2 \text{ cm}^2\text{g}^{-1}$:

$$a\ddot{\lambda}_i = 2.8 \times 10^{-5} \text{ cm s}^{-2} \qquad (4.64)$$

and

$$a\Delta\varepsilon \simeq 1.1 \times 10^4 \text{ cm} \times (\Delta t/1 \text{ day})$$
$$a\Delta I \simeq 1.1 \times 10^4 \text{ cm} \times (\Delta t/1 \text{ day}) \qquad (4.65)$$
$$a\Delta e \simeq 1.1 \times 10^4 \text{ cm} \times (\Delta t/1 \text{ day}).$$

Assuming that an orbital arc of duration $\Delta t \simeq 20$ days will be available, free from both orbital and attitude manoeuvres (see Chapter 7), at the end we get the following perturbative effects:

$$a\Delta\lambda_i \simeq \tfrac{1}{2}(a\ddot{\lambda}_i)(\tfrac{1}{2}\Delta t)^2 \simeq 100 \text{ km} \qquad a\Delta\varepsilon \simeq 2 \text{ km} \qquad (4.66)$$

and

$$a\Delta I \simeq 2 \text{ km} \qquad a\Delta e \simeq 2 \text{ km}. \qquad (4.67)$$

Thus we have estimated the total effects of the solar radiation perturbation in longitude, inclination and eccentricity; we have used very simplified formulae, since of course they should also contain a numerical factor depending on both the shape and the optical properties of the spacecraft surface. Clearly, not all the perturbation is unpredictable; a computer model can be written based on the knowledge of the spacecraft shape and optical properties. At present, such models for three-axis stabilised satellites of complex shape cannot be accurate to better than about 10%, implying that for the previous example the amount of unknown perturbative effect can be estimated to be of the order of 10 km for the longitude effects and of 200 m for the eccentricity and inclination terms.

CONCLUSIONS AND EXAMPLES

These figures must be compared with the corresponding effects of the geophysically interesting phenomena to be studied. As an example, the solar radiation perturbation in longitude masks the effect of the resonant geopotential coefficients, therefore their value cannot be determined with a relative accuracy better than the ratio between the unknown $a\,\Delta\lambda_i$ term caused by radiation pressure and the accumulated longitude effect of the resonant coefficients.

Instead of being three-axis stabilised, geostationary satellites can be one-axis stabilised; in this case they are usually made of a rapidly spinning cylinder-like body plus a small outside antenna that is despun, i.e. rotates slowly in resonance with the orbital period. For the spacecraft body the result (4.19) is valid and therefore we expect long-periodic effects in λ_i to be due to the antenna only and also to contain the misalignment angle (these small antennae are usually of simple shape). The corresponding order-of-magnitude formula is (4.37). Equation (4.44) will also be satisfied and the effect in ε can be estimated by (4.45) for the body and by (4.50) for the antenna. As for the effect in inclination, again equations (4.52) are satisfied and we can use the formulae (4.53) for the body and (4.56) for the antenna. The effect in eccentricity is given by (4.60). Using an area-to-mass ratio of $0.1\text{ cm}^2\text{g}^{-1}$, a misalignment angle of $1°$, a ratio between the antenna and the satellite cross section of $\frac{1}{5}$, an eccentricity of 0.001 and an orbital arc length $\Delta t = 30$ days we get

$$(a\ddot{\lambda}_i)_A \simeq 4.8 \times 10^{-8} \text{ cm s}^{-2}$$
$$a\Delta\varepsilon \simeq 5.4 \text{ cm} \times (\Delta t/1 \text{ day})$$
$$(a\Delta\varepsilon)_A \simeq 1.1 \times 10^3 \text{ cm} \times (\Delta t/1 \text{ day}) \qquad (4.68)$$
$$a\Delta I \simeq 5.4 \text{ cm} \times (\Delta t/1 \text{ day})$$
$$(a\Delta I)_A \simeq 1.1 \times 10^3 \text{ cm} \times (\Delta t/1 \text{ day})$$
$$a\Delta e \simeq 5.4 \text{ cm} \times (\Delta t/1 \text{ day})$$

and therefore, after 30 days, we get a longitude drift caused by the antenna of about 0.4 km (while the ε, e and I effects are much smaller). Because of the simple shape of both the body and the antenna the solar radiation perturbation can be modelled in this case to a better accuracy than in the case of three-axis stabilised satellites, say to 5%. However, this will not improve the estimated perturbation in λ_i, because this effect comes from expression (4.37), where the misalignment angle appears; the latter is not likely to be known to an accuracy of a few arcminutes. Therefore, we expect the unknown part of the solar radiation effects to be some few hundred metres in longitude, and of about 15 and 80 metres respectively for the inclination and the eccentricity terms.

Some geostationary satellites are formed by a rapidly spinning cylinder with an antenna system embedded in the satellite, so that no outside despun antenna is needed. With the same assumptions as for the spacecraft body of the previous example, at the end of the 30 days arc we shall have in this case $a\Delta\varepsilon \simeq a\Delta I \simeq 1.6$ m, $a\Delta e \simeq 1.6$ km and no contribution from λ_i. If the radiation pressure model is accurate to 5%, the unknown part of these effects will be very small (about 8 cm) for the ε and I terms, and of the order of 80 m for the eccentricity term. By comparison of these estimates with the previous ones, we can see that the situation is very much improved. In particular, the unmodelled long-periodic effect in longitude for a rapidly spinning cylinder with no despun antenna is negligible. The perturbation due to the eccentricity term is not a severe limitation. As pointed out on p.66, it does not have the same signature as the quadratic longitude effect caused by the resonant coefficients of the geopotential; therefore, in optimal experimental conditions, it is possible to separate the two effects from one another.

However, we must not forget that so far only the long-periodic effects of the perturbation have been estimated by averaging over one revolution of the satellite. These are surely the most critical effects whenever data over long orbital arcs (i.e. many satellite revolutions) are analysed. But in some cases the short-periodic effects, producing oscillations of the orbital elements over periods comparable with the orbital period, are not negligible and must be accounted for. We shall now estimate them with the same Fourier expansion technique used earlier for the long-periodic terms, by computing the coefficients with $k_2 = \pm 1$; for the semi-major axis we get

$$\dot{a}_{k_1, \pm 1} = \frac{1}{(2\pi)^2} \int_0^{2\pi} \int_0^{2\pi} \dot{a} \exp[-i(k_1 M_1 \pm M)] \, dM_1 \, dM \quad (4.69)$$

where, to zero order in e

$$\dot{a} = \frac{2}{n} T + O(e) \quad (4.70)$$

(see equation (4.9)). Using for T the expression (4.39), where, to

CONCLUSIONS AND EXAMPLES

zero order in e, f can be changed into M, we obtain

$$\dot{a}_{k_1,\pm 1} = \frac{1}{(2\pi)^2} \int_0^{2\pi} \frac{2}{n} F_y(S) \exp(-ik_1 M_1) \, dM_1$$

$$\times \int_0^{2\pi} \cos M \exp(\pm iM) \, dM - \frac{1}{(2\pi)^2} \int_0^{2\pi} \frac{2}{n} F_x(S) \exp(-ik_1 M_1) \, dM_1$$

$$\times \int_0^{2\pi} \sin M \exp(\pm iM) \, dM + O(e) \tag{4.71}$$

where the integrals in dM are not zero. In the same way it is possible to show that the antenna also contributes to short-periodic effects in a to zero order in e. Therefore, we expect to have a variation in semi-major axis with the same period as the orbital period of the satellite and a magnitude of the order of

$$(\dot{a})_{\rm sp} \simeq \frac{2}{n} \frac{\mathcal{A}\Phi_\odot}{\mathcal{M}c} \tag{4.72}$$

which would produce an oscillation of amplitude

$$(\Delta a)_{\rm sp} \simeq \frac{1}{2\pi^2} \frac{\mathcal{A}\Phi_\odot}{\mathcal{M}c} P^2. \tag{4.73}$$

From equation (4.57) it is also apparent that the eccentricity undergoes a short-periodic perturbation as well, resulting in a variation in distance $(a\Delta e)_{\rm sp}$ of about the same order as (4.73). Of course, the equations should be solved together to find out how the two effects are mixed up. However, as far as we are interested only in order-of-magnitude estimates, for instance to be compared with the accuracy of the tracking data, we can use (4.73) as a reasonable estimate. The acceleration in λ_i corresponding to $\dot{a}$ is given by (see (4.36))

$$(a\ddot{\lambda}_i)_{\rm sp} \simeq 3 \frac{\mathcal{A}\Phi_\odot}{\mathcal{M}c} \tag{4.74}$$

giving rise to an oscillation of amplitude

$$(a\Delta\lambda_i)_{\rm sp} \simeq \frac{3}{4\pi^2} \frac{\mathcal{A}\Phi_\odot}{\mathcal{M}c} P^2. \tag{4.75}$$

A short-periodic effect to zero order in e appears in ε as well (see (4.40)), caused by the R and W components of the perturbing force, whereas only the T component comes into play in equation

(4.70). By using the estimate $\mathcal{A}\Phi_\odot/\mathcal{M}c$ for the total solar radiation acceleration, we have

$$(a\dot{\varepsilon})_{sp} \simeq \frac{1}{n}\frac{\mathcal{A}\Phi_\odot}{\mathcal{M}c} \qquad (4.76)$$

and an oscillation during every orbital period of amplitude

$$(a\Delta\varepsilon)_{sp} \simeq \frac{1}{4\pi^2}\frac{\mathcal{A}\Phi_\odot}{\mathcal{M}c}P^2. \qquad (4.77)$$

In the case of a geosynchronous satellite with $\mathcal{A}/\mathcal{M} = 0.1\text{ cm}^2\text{ g}^{-1}$, the amplitude of the diurnal oscillation in semi-major axis would be

$$(\Delta a)_{sp} \simeq 17 \text{ m}. \qquad (4.78)$$

The smaller the value of $\mathcal{A}/\mathcal{M}$, the smaller will be this effect; the better the physical model used to account for it, the smaller the unknown component will be. It seems quite difficult, however, to get uncertainties smaller than one metre for satellites of complex shape.

As for the short-period effect in inclination, we use equation (4.51a) to zero order in e and its Fourier expansion, where the short-period coefficients $\dot{I}_{k_1,\pm 1}$ are easily shown not to average out. The same result can be proved by using the equations regularised for $I = 0$ in the case of satellites with small inclination and eccentricity (e.g. geostationary satellites). Therefore a diurnal effect will appear whose amplitude is of the order of

$$(a\Delta I)_{sp} \simeq \frac{1}{4\pi^2}\frac{\mathcal{A}\Phi_\odot}{\mathcal{M}c}P^2 \qquad (4.79)$$

($\mathcal{A}\Phi_\odot/\mathcal{M}c$ is again an upper limit, since only the W component contributes to change the orbital plane).

For a geostationary satellite with $\mathcal{A}/\mathcal{M} = 0.1\text{ cm}^2\text{ g}^{-1}$, the maximum amplitude of the diurnal oscillation of the orbital plane turns out to be

$$(a\Delta I)_{sp} \simeq 8.6 \text{ m}. \qquad (4.80)$$

As an example, if we are interested in the measurement of polar motion effects (due to the fact that the instantaneous rotation axis of the Earth wobbles around the axis of maximum moment of inertia with an angular amplitude of the order of 0.2 arcseconds),

CONCLUSIONS AND EXAMPLES

we must be able to 'detect' on a geosynchronous satellite an oscillation $(a\Delta I)_{pm}$ of about 40 m. The effect is diurnal, because the tracking stations rotate around the instantaneous spin axis of the Earth and therefore see the orbital plane of the satellite as oscillating with their own rotation period. Since, by equation (4.80), the short-periodic perturbative effect on the inclination cannot usually be modelled better than to an accuracy of tens of centimetres, the polar motion effect is masked by it, at least at the 1% accuracy level.

Let us now give some figures for the effects of solar radiation on a GPS satellite with a period of 12 hours. Since the purpose of spacecraft of this class is to broadcast a powerful radiowave beam, they must necessarily have large steerable solar panels and directional antennae pointing toward the Earth. As a consequence, a quadratically accumulating effect in longitude will be present, its magnitude being a significant fraction of the maximum value given by expression (4.61). With $\mathcal{A}/\mathcal{M} = 0.2 \text{ cm}^2 \text{g}^{-1}$, after five orbital revolutions an along-track displacement of about $\Lambda \times 6$ km will occur. The coefficient Λ is a complex function of the spacecraft shape and attitude, but even if it were no more than 0.1 and a radiation pressure model with a 10% relative accuracy were used, we would still be left with an unknown along-track effect of 60 metres. This means that the ephemerides of the GPS satellites need to be very frequently updated with fresh tracking data. If GPS were to be used to derive station positions accurate to the few centimetres level, simultaneous observations of more than one satellite from many stations should be carried out. Here 'simultaneous' does not mean literally 'at the same instant', but means that the observations must be spaced by spans of time small enough to keep the unmodelled perturbations in position within a few centimetres; that is not longer than a few hours.

The solar radiation effects are also very difficult to model for the SEASAT and ERS-1 radar altimeter satellites. No results such as (4.19), (4.44) or (4.37) can be proved; both the long- and short-periodic effects in λ_i and ε are of zero order in e. Because of quite a short orbital period (about 6×10^3 s, corresponding to a semi-major axis of about 7100 km) the short-periodic solar radiation effects in λ_i and ε have an amplitude of the order of 10 cm only (for $\mathcal{A}/\mathcal{M} \simeq 0.2 \text{ cm}^2 \text{g}^{-1}$). However, in less than one day (about 15 revolutions) the accumulated effect in λ_i becomes of the order of 1 km, and there is no hope of an orbit prediction to the accuracy of a few centimetres.

It is not surprising that a satellite like LAGEOS, especially devoted to an accurate determination of the interstation distances and to the recovery of polar motion parameters, has been built to a spherical shape. In this way, because of symmetry reasons and no matter how the spacecraft is spinning, the direct solar radiation force can be written as a function of the Sun's position only and equations (4.19) and (4.44) can both be proved. The area-to-mass ratio is also small ($\mathcal{A}/\mathcal{M} \simeq 0.007$ cm^2 g^{-1}), in order to provide a small non-gravitational acceleration. As an example of how the spherical shape helps in modelling the solar radiation effect, we show how the force due to the absorbed, reflected and diffused light components can easily be computed. For instance, for the diffusion we must integrate the elementary force (4.7) over the satellite hemisphere illuminated by the Sun (see figure 4.3):

$$F_\delta = -S \frac{\Phi_\odot}{c} \bar{\delta} \int_0^{\pi/2} \cos \beta \; 2\pi R^2 \sin \beta \; d\beta$$

$$-S \frac{\Phi_\odot}{c} \bar{\delta} \frac{2}{3} \int_0^{\pi/2} \cos^2 \beta \; 2\pi R^2 \sin \beta \; d\beta \qquad (4.81)$$

$$= -\frac{\Phi_\odot}{c} \bar{\delta} \pi R^2 S - \frac{\Phi_\odot}{c} \frac{2}{3} \bar{\delta} \frac{2}{3} \pi R^2 S = -\frac{13}{9} \frac{\Phi_\odot}{c} \pi R^2 \bar{\delta} S$$

where $\bar{\delta}$ is the mean diffusion coefficient of the satellite surface.

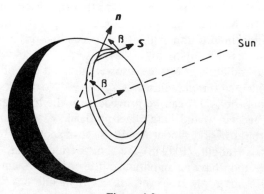

Figure 4.3

CONCLUSIONS AND EXAMPLES

Repeating the computation for absorption and reflection, we can obtain the following expression of the 'effective area' of the satellite to be used as its cross section $\mathcal{A}$:

$$\mathcal{A} = \left(\bar{\alpha} + \bar{\rho} + \frac{13}{9}\bar{\delta}\right)\pi R^2 = \left(1 + \frac{4}{9}\bar{\delta}\right)\pi R^2 \qquad (4.82)$$

(from equation (4.1)). Thus accurate ground measurements of the diffusion coefficient can be used to predict very precisely the magnitude of the perturbation.

The long-periodic effect in longitude comes from ε and is demultiplied by a factor e. Equation (4.45) yields an effect of 0.04 cm per orbital period (since $P = 1.35 \times 10^4$ s and $e = 0.004$). However, a larger long-periodic effect in eccentricity is present, since it is of zero order in e; from equation (4.60), we can estimate an $a\Delta e$ of about 9.4 cm per orbital period. Thus we get an oscillation in distance of LAGEOS over its orbital period whose amplitude changes by several centimetres per period. This can be compared with the effect of charged particle drag, which as we shall discuss in Chapters 5 and 6, produces an unmodelled decay of the semi-major axis of LAGEOS of 0.1 cm day^{-1}. Using equation (4.36) to obtain $\dot{\lambda}_i$ and integrating over one month (which is a typical arc length for the LAGEOS data analyses currently carried out), we get that drag causes an $a\Delta\lambda_i$ of about 27 metres. The effect in eccentricity appears along-track as well, and after one month it amounts to about 18 m; with a radiation pressure model accurate to 2% we are still left with an unknown solar radiation effect of 36 cm and therefore the derived values of charged particle drag can be wrong by about 1%. It is worth stressing that the solar radiation effect on the eccentricity grows linearly with time (at least on intervals significantly shorter than one year), whereas the charged drag effect grows quadratically; after about 20 days it is of the same size as the solar radiation effect. The effect in eccentricity also requires that very short orbital arcs are analysed to determine the interstation distances to a centimetric level; in less than two days the change in eccentricity will cause distance effects of the order of one metre, and therefore errors of a few centimetres even with accurate radiation pressure models.

As for the determination of polar motion parameters with LAGEOS, the accuracy is limited by the tracking technology. The solar radiation effect in inclination is of first order in e, and equations (4.55) yield in the case of LAGEOS a 0.04 cm displacement per orbital period. On the other hand, the polar motion effect appears as a diurnal oscillation of amplitude $(a\Delta I)_{pm} \simeq 12$ m, while the solar radiation effect over one day is still of order a few tenths of a centimetre. If the laser tracking data were accurate to one centimetre, polar motion could be determined to a relative accuracy of about 10^{-3}, which is better than the classical astrometric determinations by more than one order of magnitude.

As for the short-periodic effects of the solar radiation force on LAGEOS, they are of the order of a few centimetres both in longitude and in inclination (see equations (4.75, 4.77 and 4.79)); even a rough model is enough to prevent them from degrading the accuracy of the laser tracking data.

We conclude this chapter by referring back to §4.1, where possible methods of facing the problem of direct solar radiation force on artificial satellites were discussed. We have learned how to compute the order of magnitude of the solar radiation effects on the orbital elements and their typical periods; we have estimated the unmodelled portion of these effects after some numerical model has been implemented and compared it with the physical signal to be detected, in order to estimate the achievable accuracy of the quantities to be measured. Useful hints about the input data required for an accurate model also come from the theoretical analysis of the solar radiation interaction with the spacecraft. It turns out that in most cases the solar radiation perturbation is the main limitation to reaching a better accuracy in satellite geodesy experiments; tracking technologies are sometimes ahead of both our physical understanding of the non-gravitational perturbations and our capability of providing satisfactory models.

Finally, for a more rigorous presentation of the subject discussed in this chapter, we refer to the papers by Anselmo *et al* (1981, 1983a), where explicit computations of the various perturbative effects are presented and discussed in detail.

5

RADIATION PRESSURE : INDIRECT EFFECTS

Every satellite is perturbed by its interactions with the electromagnetic radiation not only via direct radiation pressure caused by the sunlight, but also in a number of more complex and subtle ways. Some of these 'indirect' effects—due to Earth-reflected radiation, thermal emission by the satellite itself, eclipses affecting portions of the orbit—have been briefly addressed in Chapter 2. In this chapter, on the other hand, we intend to present some more detailed examples of how these effects can be physically modelled and accounted for in the orbital analysis, especially in cases (like that of LAGEOS) when a reliable estimate of many tiny perturbing forces and of their long-term effects is essential to the understanding of the orbital data and also to a very precise recovery of geophysically relevant parameters (see, for example, Carpino *et al* 1986). We stress that our purpose is not to develop an exhaustive treatment of each particular effect, nor a complete list of possibly relevant phenomena for different spacecraft configurations and orbits. Instead, we hope to show the main problems that arise in the course of the modelling effort, the methods and tools which can be applied to tackle these problems and, finally, the inescapable approximations and limitations inherent to every dynamical model.

5.1 EARTH-REFLECTED RADIATION PRESSURE

After the first few years of LAGEOS orbital analysis, it became clear that the satellite's motion in the along-track direction could be predicted and/or modelled far less well than in the cross-track direction, even for arcs of a few days. This is due to the action of a drag-like force resulting in a well defined constant deceleration of about -3.4×10^{-10} cm s^{-2}, upon which a long-term modulation (with a root mean square amplitude of 1.3×10^{-10}) is superimposed. As we shall discuss in Chapter 6, the main secular part of this unpredicted perturbation is very probably connected with an amplified drag effect, due to the positive ions which populate the Earth's plasmasphere and are focused by the negative electrostatic potential acquired by the satellite. But this drag effect cannot plausibly explain several long-period harmonic components which appear in the spectrum of the unmodelled perturbative effect. Among the existing physical mechanisms which could give rise to these components, one of the most promising candidates is the radiation pressure from Earth-reflected sunlight. When one takes into account the variations of the Earth's reflective power at different latitudes and during different seasons (due to varying sea–land distribution and meteorological conditions, respectively), the resulting forces can account for many prominent lines in the low-frequency region of the spectrum of the acceleration residuals (Anselmo *et al* 1983b). The relevance of the Earth's albedo perturbative effect has been known for a long time. For instance Smith (1970) in a review paper on the subject stated: 'At . . . (these) heights, where the albedo perturbation is probably much larger than air drag and for orbits in complete sunlight, the albedo could be the major perturbing force of the semi-major axis.' But the albedo is influenced by so many geographical and meteorological variables that it is very difficult to find a model simple enough for celestial mechanical computations and realistically describing significant features of the phenomenon. For instance, it is easy to show that every albedo model symmetrical with respect to a north–south transformation cannot yield appreciable long-term perturbations on LAGEOS. On the other hand, although no albedo effect can be responsible for the strictly secular part of LAGEOS' semi-major axis decay, a number of long-periodic modulations arise when the adopted albedo model is realistic enough.

In order to derive the long-term effects of the perturbation, a suitable procedure consists of decomposing the corresponding acceleration into its R, T and W components along the radial, in-plane transverse and out-of-plane directions; then the variation of these components averaged over one revolution determines (see §§3.2 and 3.4) the short-periodic, long-periodic and secular effects. A qualitative idea of the behaviour of the R and T components can be derived from the following considerations: if we assume that the Earth's albedo has no strong large-scale variation, the R component is very roughly constant for most of the half revolution during which the satellite is over the illuminated hemisphere and is negligible in the other half; on the contrary, the acceleration component perpendicular to the plane of the terminator ('pseudo-transverse' component) gives a thrust forward when the satellite enters the dark hemisphere and a thrust backwards when it enters the illuminated hemisphere. Then the transverse T component is obtained by projecting the 'pseudo-transverse' one on the satellite's orbital plane.

The Gauss equation for the semi-major axis a (see equation (3.30)) becomes at the first order in the eccentricity

$$\dot{a} = \frac{2}{n}[T + e(R \sin M + T \cos M)] \quad (5.1)$$

where M is the mean anomaly and n is the mean motion. Clearly, for a satellite in nearly circular orbit (for LAGEOS $e = 0.004$), the long-term evolution of the semi-major axis is dominated by the average of the T component. This average would vanish if the optical properties of the Earth's surface were always equal at antipodal points, because the two opposite thrusts would be equal; in this case the dominant effects would be demultiplied by a factor e. But the sub-satellite point crosses the terminator at opposite latitudes, and there are at least two reasons why the average of T is not zero. First, the northern and southern hemispheres of the Earth have a different sea–land distribution (resulting also in a different mean albedo). Second, when it is winter in the northern hemisphere, it is summer in the southern hemisphere and *vice versa*, producing a seasonal asymmetry of cloudiness, snow cover, vegetation, etc: these factors can substantially influence the amount of reflected light. A simple model accounting for both these phenomena has been proposed by Anselmo *et al* (1983b): the average $\bar{T}$ of T over one revolution resulting from the unbalancing of the two terminator thrusts can be approximated by the following expression

$$\overline{T} = (A \cos \varphi_p \sin \lambda'_\odot + B \cos \varphi_p)\sin \varphi \quad (5.2)$$

where φ is the angle between the orbital plane and the terminator plane ($0 \leq \varphi \leq \pi$) and $\sin \varphi$ accounts for the projection of the 'pseudo-transverse' component on the orbital plane, φ_p is the co-latitude of the intersection between the orbital plane and the terminator plane (corresponding to the entrance of the satellite into the 'day') and the factor $\cos \varphi_p$ accounts for the fact that the asymmetry between the two hemispheres should be more relevant as the terminator is crossed closer to the poles, and $\lambda'_\odot = \lambda_\odot - \delta$ is a seasonal phase angle defined by the longitude $\lambda_\odot$ of the Sun and by a phase lag δ accounting for the shift of the meteorological effects with respect to the astronomical seasons. The amplitudes of the two effects described above are given by the constant coefficients B (for the mean 'hemispheric' asymmetry) and A (for the seasonal variation). An order-of-magnitude estimate of these coefficients can be derived as follows. At LAGEOS' altitude of about 6000 km, the maximum value of the R component produced by the albedo effect is about 10% of the acceleration due to direct solar radiation pressure, i.e. about 3.7×10^{-8} cm s^{-2}. The peak value of the pseudo-transverse component is about ten times smaller (see Levin 1962); the integrated effect of each thrust is probably no more than 15% of the peak value, and if we assume that there is a 10% asymmetry between the two hemispheres and a 20% seasonal variation, we get $|B| \simeq 5.6 \times 10^{-11}$ and $|A| \simeq 1.1 \times 10^{-10}$ cm s^{-2}. It appears remarkable that these estimates are in reasonable agreement with the magnitude of the long-term components of the unmodelled along-track acceleration of LAGEOS. As for the signs of A and B, if the southern hemisphere is on the average brighter, the mean albedo increases in winter, and the phase lag δ is small, then $B, A > 0$. Now we have to compute the orbital effects of this perturbing force and to show that the model gives the required long-periodic terms.

The satellite's inclination (I) and longitude of the ascending node (Ω), defined as in Chapter 3, allow us to derive, in the equatorial reference frame, the unit vector N perpendicular to the orbital plane

$$N = [\sin I \sin \Omega, -\sin I \cos \Omega, \cos I]. \quad (5.3)$$

The direction of the Sun is specified by the unit vector S, which, in an equatorial reference frame (see figure 3.1) with the x-axis pointing towards the vernal equinox, is given by

$$S = [\cos \lambda_\odot, \sin \lambda_\odot \cos \varepsilon_\odot, \sin \lambda_\odot \sin \varepsilon_\odot] \quad (5.4)$$

where $\varepsilon_\odot$ is the obliquity of the ecliptic. The intersection line between the orbital plane and the terminator plane is defined by the vector product $S \times N$; hence the co-latitude φ_p at which the orbit enters the 'day' half-space is given by

$$\cos \varphi_p = z \cdot (S \times N)/|S \times N| \quad (5.5)$$

where the unit vector z points towards the north pole. The angle φ between the two planes is given by

$$\sin \varphi = |S \times N|. \quad (5.6)$$

Substituting in equation (5.2) we find for the averaged T component

$$\bar{T} = [z \cdot (S \times N)](A \sin \lambda_\odot + B). \quad (5.7)$$

The triple product can be computed from equations (5.3) and (5.4):

$$z \cdot (S \times N) = -\sin I[\cos^2(\tfrac{1}{2}\varepsilon_\odot)\cos(\lambda_\odot - \Omega) \\ + \sin^2(\tfrac{1}{2}\varepsilon_\odot)\cos(\lambda_\odot + \Omega)]. \quad (5.8)$$

Now we can directly decompose $\bar{T}$ into Fourier components

$$\bar{T} = -\tfrac{1}{2} A \sin I \cos^2(\tfrac{1}{2}\varepsilon_\odot)\sin(2\lambda_\odot - \Omega - \delta) \\ - \tfrac{1}{2} A \sin I \sin^2(\tfrac{1}{2}\varepsilon_\odot)\sin(2\lambda_\odot + \Omega - \delta) \\ - B \sin I \cos^2(\tfrac{1}{2}\varepsilon_\odot)\cos(\lambda_\odot - \Omega) \\ - B \sin I \sin^2(\tfrac{1}{2}\varepsilon_\odot)\cos(\lambda_\odot + \Omega) \\ - \tfrac{1}{2} A \sin I (1 - \cos^2 \delta \sin^2 \varepsilon_\odot)^{1/2} \sin(\Omega + \delta') \quad (5.9)$$

where the phase angle of the Ω harmonic is

$$\delta' = \arctan\left(-\frac{\tan \delta}{\cos \varepsilon_\odot}\right) \simeq -\delta. \quad (5.10)$$

In order to compute the frequencies of the various terms we neglect the eccentricity of the Sun, so that $\dot{\lambda}_\odot$ can be considered constant, and recall that for LAGEOS $I = 110°$, $\dot{\Omega} = 0°.343$ per day.

Table 5.1 summarises the relative amplitudes and the phases of the resulting five long-periodic components of the T acceleration. The effects on the semi-major axis are obtained simply by equation (5.1): long-periodic perturbations do appear as a direct consequence of the assumptions implicit in equation (5.2), i.e. of the north–south asymmetric albedo model. We have to note that neglecting entirely the effects of the R component of the acceleration is not completely correct: although demultiplied by a factor e, these effects are not negligible because the average radial component is much larger than the transverse one, since both the diffused and the reflected light components contribute to it for about half of the orbit. Computing the Fourier expansion of the corresponding semi-major axis variations is not difficult, provided one chooses a simple approximate 'model' of R and then follows the same procedure we have outlined for the T component (we refer for the details of the computations to the paper by Anselmo *et al* (1983b)). The results show in fact that, for the eccentricity of LAGEOS, the largest Fourier components of the R perturbation are smaller by about one order of magnitude than the largest T components.

Table 5.1 Fourier components of the T acceleration due to Earth-reflected sunlight on the orbit of LAGEOS ($\dot{a} \simeq 2T/n$).

Argument	Period (days)	Amplitude	Phase for terms: amplitude∗ sin(argument + phase)
$\lambda_\odot - \Omega$	560	$0.90\,B$	$3\pi/2$
$\lambda_\odot + \Omega$	271	$0.04\,B$	$3\pi/2$
$2\lambda_\odot - \Omega$	221	$0.45\,A$	$\pi - \delta$
$2\lambda_\odot + \Omega$	156	$0.02\,A$	$\pi - \delta$
Ω	1050	$0.47(1 - 0.16\cos^2\delta)^{1/2}A$	$\pi - \arctan(-0.92\tan\delta)$

Inspection of table 5.1 shows another interesting feature: the frequencies appearing in equation (5.9) are the same as those which characterise the solar-tidal effects on the orbit (the main terms in table 5.1 have the periods of the so-called S_1, P_1 and K_1 tides; see Lambeck (1980, Chap. 6)). This coincidence had to be expected because the frequencies of both phenomena depend only on the geometric configuration of the orbit with respect to the Sun's position. As shown by Barlier *et al* (1986), for LAGEOS the magnitude of the two effects is also comparable. However, it does not seem plausible that the tidal effects are a major source of the long-periodic unmodelled perturbations, because a refined model of them was used in the orbit determination procedure, and the 'residual' uncertainty is smaller by at least an order of magnitude than the albedo perturbations (according to our previous estimates; we recall that the asymmetric albedo effect was not included at all in the dynamical models used to analyse LAGEOS data).

The conclusion is that earthshine radiation pressure must be a relevant factor in the long-period perturbations on LAGEOS' semi-major axis. However, the problem arises of assessing how well a semiquantitative model like (5.2), based upon the simplistic idea of two unbalanced terminator thrusts, describes the physical reality of the perturbing force due to light scattered by the Earth. The main problem lies in the complexity of the phenomenon of the light scattering by a variable planetary surface and atmosphere. The way in which this process occurs can be the crucial factor in determining whether or not long-periodic effects arise. As an example, we can give the following result about the diffused component of earthshine. Let us assume that

(*a*) the radiation coming from the Earth's surface is the sum of two components, one ('diffused light') depending only on the point P on the Earth's surface and on the satellite's zenith angle, i.e. the angle ζ between the satellite's direction (specified by the unit vector s) and the normal to the diffusing surface element $dS(P)$ (unit vector n); the other ('reflected light') depending also on other variables, e.g. the Sun's direction.

(*b*) The Earth is assumed to be spherical and to have optical properties independent of longitude (so that Earth's rotation does not matter).

(*c*) The radiation pressure acting on a satellite is assumed to be proportional to the light intensity $\mathscr{L}(\zeta, P)\, dS(P)$, to be directed along s and to depend on the distance D between the surface element and the satellite, but not on the satellite's attitude.

(*d*) The satellite's orbit is assumed to be circular.

Then the tangential component of the perturbing force on the satellite caused by radiation pressure from diffused light has zero average over one orbital period, and thus cannot produce any long-term orbital effect.

This theorem can easily be proved as follows (see Barlier *et al* 1986). The tangential component of the perturbing force, averaged over one orbital revolution, will be proportional to the integral

$$\int_0^{2\pi} d\lambda \int \frac{\mathscr{L} \cos \gamma}{D^2} \, dS(P) = \int K(P) \, dS(P) \qquad (5.11)$$

where $K(P)$ is the function to be integrated on the illuminated Earth's surface, the anomaly λ gives the satellite's position along the orbit and γ is the angle between s and the orbital velocity. Let us define the new variable $\bar{\lambda} = \lambda - \lambda_0$, where λ_0 is the anomaly of the point on the orbit closest to P (see figure 5.1). Then we have

$$K(P) = \int_0^{2\pi} \frac{\mathscr{L}(\zeta(\lambda, P), P) \cos \gamma(\lambda, P)}{D^2(\lambda, P)} \, d\lambda$$

$$= \int_{-\pi}^{\pi} \frac{\mathscr{L}(\zeta(\bar{\lambda}, P), P) \cos \gamma(\bar{\lambda}, P)}{D^2(\bar{\lambda}, P)} \, d\bar{\lambda}. \qquad (5.12)$$

In the right-hand side of equation (5.12), the functions D, ζ and I are even with respect to the variable $\bar{\lambda}$, whilst $\cos \gamma$ is odd (since the orbit is circular and therefore $\gamma(\bar{\lambda}) = \pi - \gamma(-\bar{\lambda})$). As a consequence, $K(P) = 0$ for every P and the average tangential acceleration produced by the diffused light is zero. Note that the basic assumptions of this theorem are quite reasonable in the case of LAGEOS, which has an almost spherical shape and a very low eccentricity.

As a result of the zero effect of the diffused light, reflected light comes to the fore, although the total amount of reflected light from the whole surface of the earth is less than that of diffused light by about a factor of four. The reflective properties of the Earth's surface can be inferred, for example, by the diurnal variations of Earth albedo values obtained by Meteosat data (see Gube 1982), which show three main features:

(*a*) the sea is usually dark, but its albedo increases by about a factor of three when the solar zenith angle increases from 0 to 70° (due to Fresnel-type reflection phenomena of the 'rough' water surface);

(*b*) a similar, if less marked, trend appears when a 'dark' land surface is seen through a cloud-free atmosphere or when optically thin clouds are present;

EARTH-REFLECTED RADIATION PRESSURE

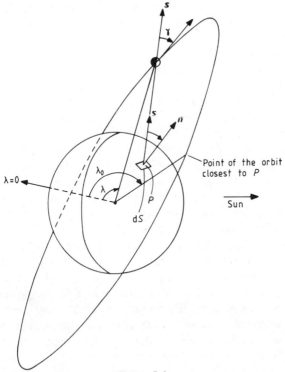

Figure 5.1

(c) in the case of optically thick clouds the albedo is high (about 0.6), but the diffused radiation predominates and the dependence on solar zenith angle practically vanishes.

These facts show that the amount of reflected light depends strongly on the surface type and on the prevailing meteorological conditions (that is, for instance, on the season), and is possibly anticorrelated with the albedo itself (i.e. dark surfaces are more likely to behave as strong reflectors). Moreover, since reflection phenomena appear to dominate for high zenith angles of the Sun, they can influence the motion of a satellite mainly when the satellite 'sees' a crescent-like Earth (i.e. mainly in the two portions of the orbit where the satellite is about to go in or has just got out of the Earth's shadow). In these conditions, the corresponding force has, unavoidably, a significant tangential component, which does not average out if there is not perfect symmetry between the reflective properties of the surface in the northern and the southern hemispheres.

The qualitative features of this effect are no different from those of the model discussed earlier—i.e. the long-term perturbative effect arises as a consequence of two unbalanced forces acting in parts of the orbit at opposite latitudes. However, a more detailed analysis of the phenomenon can no longer be based upon an explicit analytic model, but requires a more complex semi-numerical approach. Barlier *et al* (1986) have worked out such a model, accounting for: (*a*) the geometry of the reflected light paths; (*b*) the reflective properties of water surfaces; (*c*) the different sea–land distribution in the two hemispheres; (*d*) the seasonally changing, latitude-dependent cloud coverage. The along-track perturbative force has been numerically computed as a function of the three arguments λ, $\lambda_\odot$ and Ω (plus a few free parameters playing the same role as A, B and δ), averaged over λ and finally decomposed into two-dimensional Fourier harmonics of $\lambda_\odot$ and Ω. As could be expected, the dominant Fourier components of the perturbation are the same as given by the simpler analytical model, even if the relative amplitudes are somewhat changed and higher harmonics appear (such as the $(2\lambda_\odot - 2\Omega)$ one, which is clearly present in the data). The magnitude of the long-periodic acceleration is still of the order of 10^{-10} cm s^{-2}, i.e. consistent with the observed acceleration residuals.

We shall not discuss here in detail another radiation pressure effect depending on the presence of the Earth, i.e. that produced by infrared thermal radiation emitted by the planet. Thermal radiation accounts for more than 60% of the total energy flux coming from the Earth, and thus at first sight it would appear that this mechanism could give rise to effects comparable to or even larger than those due to albedo. In fact this does not occur because the Earth emits thermal radiation in a way much more symmetrical than for visible light: for instance the latter obviously has a strong day–night asymmetry, while the corresponding asymmetry in the thermal flux is only a few per cent (Raschke and Bandeen 1970). Clearly, were the outgoing radiation spherically symmetric and uniform, it would only cause a small, satellite-dependent (because of different area-to-mass ratios and optical properties) change of the 'effective' gravitational constant. To a better approximation, one can consider a latitudinal dependence of the thermal radiation field given by a spherical harmonic expansion, the main contribution obviously being due to the zonal term of second degree. The corresponding perturbative effects have been computed by Manakov (1977) and Sehnal (1981), by using simple assumptions about the emission law by Earth's elementary areas. These authors have shown that (as could be expected by analogy with the gravitational harmonics) there is no long-periodic effect on the satellite semi-major axis, eccentricity and inclination of zero order in the eccentricity, while ω and Ω experience secular changes analogous (although much smaller) to those caused by the gravitational J_{20}. For LAGEOS, Sehnal (1981) estimated that $\dot{a} \sim 10^{-6}$ cm day^{-1}, while $\dot{\omega} \sim \dot{\Omega} \sim 10^{-7}$ deg day^{-1}. Annual periodicities could also arise from seasonal changes of the odd thermal harmonics.

It is very probable that the effects discussed in this section will prove not to be the only ones contributing to the unmodelled perturbations of the LAGEOS orbit. We have already quoted the fact that a peculiar drag-like mechanism is also working (see Chapter 6), and it is likely that the corresponding deceleration is time dependent (notice that if the Sun controls in some way the drag intensity, the same periods that are listed in table 5.1 will arise again). Moreover, at the 10^{-11} cm s^{-2} level, other very small non-gravitational forces, completely negligible in ordinary conditions, could have some influence, and we shall discuss an interesting example of that in §5.2. At any rate, the asymmetric albedo effect is, in our opinion, intriguing because it shows clearly both the advantages of developing at least some empirical, semiquantitative dynamical model by the classical tools of celestial mechanics, and the very difficult problems which must be solved to substantially improve the attainable accuracy by the next generation of geophysical satellites.

5.2 ANISOTROPIC THERMAL EMISSION

Another indirect effect of the interaction between the solar radiation and an artificial satellite is due to the fact that the equilibrium temperature distribution on the satellite itself becomes non-uniform, owing to different orientations with respect to solar heating of different parts of the spacecraft body. This causes a net force to act, since the thermal photons emitted from the hotter areas of the surface carry away more momentum than those coming from colder areas. For a spinning satellite, there will be two main asymmetries of the temperature distribution. Representing the satellite as a small 'planet', we can speak of a 'seasonal' asymmetry which arises from the fact that the angle ξ between the spin axis and the Sun's direction is not $90°$ and changes with a yearly periodicity, causing a different radiation flux on the 'northern' and 'southern' sides of the satellite; on the other hand, a 'diurnal' asymmetry between the 'day' and the 'night' sides is also present whenever ξ is not zero. Moreover a finite thermal inertia will cause a time lag between the hottest and the sub-solar points (these latter effects are clearly negligible when the rotation is fast). In the following, we shall analyse in some detail the temperature distribution and the force due to anisotropic thermal emission for a rapidly spinning, nearly spherical orbiting body. This will represent quite realistically the situation for LAGEOS and like satellites; then we shall briefly discuss the complications arising for spacecraft having a more complex shape and/or thermal system.

The energy flux absorbed by the satellite's surface elements is $\alpha \Phi_\odot \mathbf{n} \cdot \mathbf{S}$, where α is the absorption coefficient, $\Phi_\odot$ is the solar constant ($\simeq 1.38 \times 10^6 \, \text{erg} \, \text{cm}^{-2} \, \text{s}^{-1}$) and the two unit vectors $\mathbf{n}$ and $\mathbf{S}$ specify the directions orthogonal (outwards) to the surface element and pointing towards the Sun, respectively. On the other hand, the energy emitted per unit area is $\varepsilon \sigma \mathcal{T}^4$, where $\sigma = 5.67 \times 10^{-5} \, \text{erg} \, \text{cm}^{-2} \, \text{s}^{-1} \, \text{K}^{-1}$ is the Stephan–Boltzmann constant, $\mathcal{T}$ is the surface temperature and ε is the emissivity coefficient ($\varepsilon = 1$ for a black body). Let us now consider the thermal state of a spherical, homogeneous and uniformly rotating body which is illuminated by the Sun. By introducing a reference frame whose z-axis is directed along the spin axis of the body (specified by the unit vector $\mathbf{w}$), while the y-axis contains the projection of $\mathbf{S}$ on the xy-plane, and using polar coordinates (r, θ, φ), we obtain

$$\mathbf{n} = [\sin \theta \cos \varphi, \sin \theta \sin \varphi, \cos \theta]$$

$$\mathbf{S} = [0, \sin \xi, \cos \xi] \qquad (5.13)$$

$$\mathbf{n} \cdot \mathbf{S} = \sin \theta \sin \varphi \sin \xi + \cos \theta \cos \xi.$$

If the rotation is rapid enough for the distribution of temperature

ANISOTROPIC THERMAL EMISSION

to reach a steady, axisymmetrical state (not depending on the time t and the azimuthal angle φ), the heat conduction equation takes Laplace's form (see e.g. Kittel and Kroemer 1980, Chap.15):

$$\nabla^2 \mathscr{T} = 0 \qquad (5.14)$$

and its general solution is an axially symmetric harmonic function, which can be expanded as

$$\mathscr{T}(r, \theta) = \mathscr{T}_0 + \sum_{n=1}^{\infty} \mathscr{T}_n \left(\frac{r}{\mathscr{R}}\right)^n P_n(\cos \theta) \qquad (5.15)$$

where $\mathscr{T}_0$ is the average temperature, $\mathscr{R}$ is the radius of the sphere, $\mathscr{T}_n$ are constant coefficients and P_n are the Legendre polynomials of order n. As a boundary condition, we have to use the balance between the outwards heat flow caused by thermal conduction, which is given by $-\chi \partial \mathscr{T}/\partial r$ (χ is the thermal conductivity of the body, assumed to be constant), and the emission at the surface (i.e. the difference between $\varepsilon \sigma \mathscr{T}^4$ and the energy flux absorbed from the Sun $\alpha \boldsymbol{n} \cdot \boldsymbol{S} \Phi_\odot$):

$$\varepsilon \sigma \mathscr{T}^4 - \alpha \boldsymbol{n} \cdot \boldsymbol{S} \Phi_\odot = -\chi \frac{\partial \mathscr{T}}{\partial r}. \qquad (5.16)$$

But the requirement of axial symmetry imposes that in equation (5.16) we replace $\boldsymbol{n} \cdot \boldsymbol{S}$ with its angular average

$$s(\theta) = \frac{1}{2\pi} \int \boldsymbol{n} \cdot \boldsymbol{S} \, d\varphi \qquad (5.17)$$

the integral being extended only to the illuminated hemisphere, i.e. for $\boldsymbol{n} \cdot \boldsymbol{S} > 0$. If we choose $\xi \leqslant \pi/2$, the analytic expression of $s(\theta)$ is

$$s(\theta) = \cos \xi \cos \theta \qquad \text{for} \quad \theta \leqslant \frac{\pi}{2} - \xi$$

$$s(\theta) = \frac{\sin \xi \sin \theta \cos \varphi_1}{\pi} + \frac{\cos \xi \cos \theta}{2\pi}(\pi - 2\varphi_1)$$

$$\text{for} \quad \frac{\pi}{2} - \xi \leqslant \theta \leqslant \frac{\pi}{2} + \xi \qquad (5.18)$$

$$s(\theta) = 0 \qquad \text{for} \quad \theta \geqslant \frac{\pi}{2} + \xi$$

where φ_1 is the solution of $\mathbf{n} \cdot \mathbf{S} = 0$, i.e.

$$\varphi_1 = -\arcsin(\cot \theta \cot \xi). \tag{5.19}$$

For $\xi > \pi/2$, $s(\theta)$ can de derived by noting that when ξ becomes $(\pi - \xi)$, $s(\theta)$ is changed into $s(\pi - \theta)$. Now we can rewrite equation (5.16) by introducing $s(\theta)$ and (provided the temperature variations are small with respect to $\mathcal{T}_0$) by linearising

$$\varepsilon\left(\sigma \mathcal{T}_0^4 + 4\sigma \mathcal{T}_0^3 \sum_n \mathcal{T}_n P_n(\cos \theta)\right) - \alpha s(\theta)\Phi_\odot = \frac{\chi}{\mathcal{R}} \sum_n n\mathcal{T}_n P_n(\cos \theta). \tag{5.20}$$

We can expand $s(\theta)$ too in a series of Legendre polynomials

$$s(\theta) = \sum_{n=0}^{\infty} s_n P_n(\cos \theta) \tag{5.21}$$

so that equation (5.20) can be separated into the following relationships:

$$\begin{aligned}\varepsilon\sigma\mathcal{T}_0^4 &= \alpha s_0 \Phi_\odot \\ \mathcal{T}_n &= \alpha s_n \Phi_\odot/(4\varepsilon\sigma\mathcal{T}_0^3 + n\chi/\mathcal{R}).\end{aligned} \tag{5.22}$$

As particular cases, we get from (5.18) $s_0 = 1/4$, $s_1 = \frac{1}{2} \cos \xi$; note that s_0 can also be simply deduced from the conservation of energy. We can rewrite equation (5.22) in the form

$$\begin{aligned}4\sigma\mathcal{T}_0^4 &= \alpha\Phi_\odot/\varepsilon \\ \mathcal{T}_n &= s_n \mathcal{T}_0/(1 + n\chi\mathcal{T}_0/\alpha\mathcal{R}\Phi_\odot).\end{aligned} \tag{5.23}$$

If we assume $\alpha = \varepsilon$, the average temperature of the body is $\mathcal{T}_0 = 280$ K.

In order to derive the force due to anisotropic thermal emission, we have to note that each surface element emits in the direction $\mathbf{n}$ a momentum flux equal to $2\varepsilon\sigma\mathcal{T}^4/3c$, where c is the velocity of light and the factor 2/3 allows for the fact that the emission is assumed to follow Lambert's law (see §4.1, where the same factor 2/3 has been derived for a Lambert diffuser). This corresponds to a net force acting on the satellite given by $-\mathbf{n}(2\varepsilon\sigma\mathcal{T}^4 \, dS/3c)$. The total force $\mathbf{F}$ can be obtained by integrating over the whole body surface: owing to the axial symmetry of the problem, it is directed along $\mathbf{w}$.

The explicit computation yields

$$F = -w\frac{2\varepsilon\sigma}{3c}\int \cos\theta\left(\mathcal{T}_0^4 + 4\mathcal{T}_0^3 \sum_n \mathcal{T}_n P_n(\cos\theta)\right) dS$$

$$= -w\frac{4\pi\varepsilon\sigma\mathcal{R}^2}{3c}\int_0^\pi \sin\theta\cos\theta\left(\mathcal{T}_0^4 + 4\mathcal{T}_0^3 \sum_n \mathcal{T}_n P_n(\cos\theta)\right) d\theta$$

$$= -w(32\pi\mathcal{R}^2\varepsilon\sigma\mathcal{T}_0^3\mathcal{T}_1/9c) \qquad (5.24)$$

or, by substituting $\mathcal{T}_1$ from equation (5.23),

$$F = -w(4\pi\alpha\Phi_\odot\mathcal{R}^2 \cos\xi/9c\beta) \qquad (5.25)$$

where β is a reduction coefficient given by

$$\beta = 1 + \frac{\chi\mathcal{T}_0}{\alpha\mathcal{R}\Phi_\odot}. \qquad (5.26)$$

Notice that since the linearised equation (5.20) is valid only provided $|\mathcal{T}_n| \ll \mathcal{T}_0$, this implies $\beta \gg 1$ (i.e. high conductivity). By a similar, straightforward procedure it is possible to show that for a stratified satellite, having a 'core' and a 'mantle' with different conductivities, the same formula (5.25) holds provided the coefficient β is re-adjusted. Some order-of-magnitude estimates can be derived by assuming that we deal with a sphere with a radius of 30 cm, a mass of 400 kg (which correspond to the properties of LAGEOS) and $\alpha = \varepsilon = 0.4$, typical values for an unpolished metal surface. In this case equation (5.26) yields an acceleration A_w given by

$$A_w = -w\frac{5.8 \times 10^{-8}}{\beta}\cos\xi \text{ cm s}^{-2}. \qquad (5.27)$$

Typical values of β are 471 for a homogeneous aluminimum sphere ($\chi = 2.1 \times 10^7 \text{ erg cm}^{-1}\text{s}^{-1}\text{K}^{-1}$), 79 for a homogeneous leaden sphere ($\chi = 3.5 \times 10^6 \text{ erg cm}^{-1}\text{s}^{-1}\text{K}^{-1}$) and 155 for a sphere with an insulating ($\chi = 0$) core of radius 25 cm and an aluminium outer shell. The order of magnitude of the temperature variations on the surface can be estimated if we call $\Delta\mathcal{T}$ the temperature difference between the north and south poles of the satellite produced by the P_1 term in equations (5.15) and (5.23). We get $\Delta\mathcal{T} = \mathcal{T}_0\cos\xi/\beta$ and since $\mathcal{T}_0 = 280$ K, for the three previous examples we obtain $\mathcal{T}_0/\beta = 0.6$, 3.5 and 1.8 K respectively.

We have shown that the acceleration due to the 'seasonal' temperature asymmetry is directed along the satellite's spin axis, with A_w function of the angle between the spin axis itself (w) and the Sun's direction (S). If w is assumed to be constant, i.e. if the satellite is rotating around the axis of maximum moment of inertia and no free or forced precession is present, and if there is no eclipse (see §5.4), then A_w is a function only of the variable S, and the results proved in §§4.2 and 4.3 for the solar radiation pressure do apply: the perturbation causes no long-periodic or secular effect on the semi-major axis and only effects of first order in the eccentricity on the satellite's longitude, inclination and node. Thus, provided arcs not including eclipses are analysed, for most orbit determination purposes the perturbation due to this effect is in fact demultiplied by a factor e; and by using equation (5.27) we can exclude the possibility that the unmodelled semi-major axis variations of LAGEOS are caused, to a large extent, by this mechanism, unless for LAGEOS the average thermal conductivity is much lower than we have assumed.

In the previous treatment we have neglected entirely the contribution of the Earth's albedo and infrared radiation to the thermal state of the satellite. The corresponding force is again directed mostly along w, but its magnitude and sign are determined by the geometric relationship between the spin axis and the orbital plane; in particular the semi-major axis effect should be maximum when w lies in the orbital plane, but should average out for a circular orbit and an axisymmetric terrestrial radiation field (as can be easily inferred by symmetry considerations). Therefore only the albedo radiation, which is strongly dependent on longitude, should give a significant contribution. But the average energy flux due to Earth's albedo is only of the order of 5% of the solar constant, and thus by equations (5.25) and (5.27) a plausible upper limit for this perturbation in the case of LAGEOS is 10^{-11} cm s^{-2}. This is again too small to account for the unmodelled terms, although is not completely negligible.

Another interesting phenomenon is the 'diurnal' temperature asymmetry we quoted at the beginning of this section, which causes a force perpendicular to the spin axis. In the case of LAGEOS, whose spin rate should be rapid (about 10 rad s^{-1} according to Rubincam (1982)), Barlier *et al* (1986) find by order-of-magnitude arguments that the maximum temperature difference between the 'day' and 'night' sides of the satellite cannot exceed 10^{-2} K, with a phase lag close to 45°. Since the force is proportional to the temperature difference, it is very small in comparison with that arising from the 'seasonal' effect, and therefore this effect can be neglected. It is important to note that this conclusion would not be valid for a slowly spinning or a three-axis stabilised spacecraft; however, in these cases the satellite's thermal behaviour is likely to be very different due to the complex shape and the 'active' thermal system, and in order to predict the temperature distribution in a number of different conditions, highly refined numerical thermal models are currently employed. At any rate, when it is known that for a satellite a 'typical' temperature difference $\Delta \mathcal{T}$ is present between two substantial parts or sides of the body, the order of magnitude of the corresponding perturbative acceleration can be evaluated from the expression

$$A_w \simeq \nu \alpha \frac{\mathcal{A}}{\mathcal{M}} \frac{\Phi_\odot}{c} \frac{\Delta \mathcal{T}}{\mathcal{T}_0} \qquad (5.28)$$

(see equations (5.23) and (5.24)), where we have introduced $\mathcal{A}/\mathcal{M}$, a 'mean' value of the area-to-mass ratio of the satellite, and a numerical coefficient ν of order unity which replaces the factor 4/9 derived for the spherical satellite.

5.3 RADIOWAVE BEAMS

Let us consider the case of a satellite which is not completely 'passive' like LAGEOS or Starlette, but is equipped with a telecommunication system capable of emitting radiowave beams towards the Earth. This will obviously cause a radiation pressure 'recoil', whose order of magnitude is given by the solar radiation pressure on the satellite demultiplied by a factor giving the fractional efficiency of the spacecraft in converting the impinging solar energy into radiowaves (see Chapter 2). The corresponding perturbative effects are easily evaluated by the techniques developed in Chapter 4, since the transmission can be assumed to occur along the unit vector b giving also the symmetry axis of the emitting antenna. This introduces a contribution to the term in equation (4.20) given by $\mathcal{W}(t)/\mathcal{M}c$, where $\mathcal{W}(t)$ is the emitted power as a function of time. This causes:

(a) An $\dot{a}$ term given by $-2\bar{\xi}\mathcal{W}(t)/n\mathcal{M}c$ (see equation (4.25)) resulting—via equation (4.35)—into a truly secular λ_i perturbation (as opposed to the essentially annual long-term effects of the solar radiation pressure) given by

$$a\Delta\lambda_i = -\frac{3}{2} \frac{\bar{\xi}\bar{\mathcal{W}}}{\mathcal{M}c} t^2 \tag{5.29}$$

where $\bar{\mathcal{W}}$ is the average emitted power and $\bar{\xi}$ is the antenna misalignment angle (measured on the orbital plane) with respect to the Earth's centre.

(b) A secular $\dot{\varepsilon}$ term given by $2\bar{\mathcal{W}}(t)/na\mathcal{M}c$, i.e. not demultiplied by the (supposedly) small $\bar{\xi}$ parameter (see equation (4.48)). This produces a longitude effect

$$a\Delta\varepsilon = -2 \frac{\bar{\mathcal{W}}}{n\mathcal{M}c} t \tag{5.30}$$

which is larger than the λ_i effect (5.29) only for a time interval less than $4/3n\bar{\xi}$, i.e. if $\bar{\xi} \simeq 1°$, about 10 orbital periods. For a GPS-like spacecraft, with $\mathcal{M} \simeq 500$ kg, $\bar{\mathcal{W}} \simeq 400$ W and an orbital period of 12 hours, after 1 day we get an along-track displacement of about 3 metres, that becomes 50 metres after 10 days. An accurate modelling of this perturbation seems quite unlikely.

The effects due to radiowave transmission on inclination and eccentricity are demultiplied by small parameters (η and e, respectively). As a consequence, these effects are usually smaller than those of zero order due to direct solar radiation, which are always present for the eccentricity and are caused by the antenna for the inclination.

5.4 ECLIPSES

When a satellite enters the Earth's shadow, its interaction with the solar radiation is suddenly turned off, and this fact obviously influences the orbital evolution. In order to model this effect the crucial point is the determination of the points of intersection of the shadow with the orbit of the satellite. If the shadow had a perfectly cylindrical shape, the satellite's eccentric anomaly (u) at shadow entry and exit could be determined by solving a quartic equation in $\sin u$ or $\cos u$. Alternatively, one can easily establish some 'shadow criterion' permitting a step-by-step search of the shadow boundaries along the orbit (Aksnes 1976). But the main problem is that in reality there is no discontinuity, because the finite size of the solar disc and the presence of the Earth's atmosphere give rise to a penumbra region, and the force due to solar radiation pressure decreases smoothly to zero from the outer to the inner edge of the penumbra. Were this force variation exactly linear, the penumbra effect would be automatically accounted for by referring the shadow limits to the centre of the Sun. But the real situation is much more complex, not only for geometrical reasons, but also because clouds and refraction in the atmosphere can substantially change the properties of the penumbra, and these somewhat erratic phenomena are very difficult to model and/or to predict. As we shall see, when an accurate orbit determination is needed, this problem can be handled in only two ways: either by excluding *a priori* from the data analysis orbital arcs including eclipses, or by using orbits with low eccentricity, for which the long-period eclipse effects on the semi-major axis are of first order in e.

A semi-analytic method to treat the eclipse perturbation has been developed by Kozai (1961) and Aksnes (1976), on the assumption that the radiation pressure force always acts along the Sun–satellite line. We shall illustrate here the general method by giving Kozai's equation for the semi-major axis change, which is of crucial importance because of its cumulative effect on the mean anomaly:

$$\Delta a = \frac{2}{n^2} \frac{\mathcal{A}}{\mathcal{M}} \frac{\Phi_\odot}{c} \left(\frac{a_\odot}{r_\odot}\right)^2 [R(0) \cos u + T(0)(1-e^2)^{1/2} \sin u]_{u_1}^{u_2}.$$
(5.31)

This equation gives in general the whole (short-period) perturbative effect during an orbital arc which is entirely in sunlight and spans the range u_1 to u_2 in eccentric anomaly; $R(0)$ and $T(0)$ are here the R and T components of the radiation pressure acceleration evaluated at zero true anomaly, i.e. at perigee. Note that, of course, the order of magnitude of the perturbation is consistent with that given by equation (4.73). If we now identify u_1 and u_2 with the eccentric anomalies at two consecutive shadow exits and entries, it is easily seen that in general the short-period effects at the two shadow boundaries will not balance out exactly, and this 'residual' displacement will produce long-period terms (the long-period arguments will contain the angle $2(\lambda_\odot - \Omega)$, which controls the orientation of the orbital plane with respect to the Sun and as a consequence the periodicity of eclipses). From equation (5.31) we can derive several interesting conclusions:

(a) The eclipse effect on the semi-major axis is of first order in e. This can be seen by analysing the case of a circular orbit: if we measure the true anomaly from the projection of the Sun's direction on the orbital plane, $T(0)$ is zero, $u_1 = -u_2$ and, since cos u is even, Δa is zero. (A more intuitive way of stating this argument is: the force is directed along S, the shadow is symmetrical about S, therefore for symmetry reasons any angular average of the T component vanishes for a circular orbit.)

(b) For a revolution including an eclipse, the order of magnitude of the semi-major axis effect is given by the eccentricity (as proved above) times the amplitude of the short-periodic radiation pressure effect multiplied by $(u_2 - u_1)/2\pi$. Thus in the case of LAGEOS, since the short-periodic effect is of about 1.4 cm (according to the estimates of Chapter 4), $e = 0.004$ and at an altitude of about two Earth radii the satellite spends about 1/15 of the time in shadow, eclipses cause a long-period semi-major axis drift of the order of 2×10^{-3} cm day^{-1}. Owing to the small eccentricity of the orbit, this effect is well below the unmodelled observed effects quoted in previous sections.

(c) The uncertainty connected with the poor modelling of the penumbra can be assessed by estimating that u_1 and u_2 can be affected by an error ranging from $0°.1$ to $1°$, due to the finite diameter of the Sun and to the atmosphere's angular thickness (which depends on the satellite's height). Thus when the orbit includes eclipses it seems very difficult to avoid errors of the order of 10^{-3} of the short-term radiation pressure effect per revolution for orbital elements like the nodal longitude or the inclination, for which perturbative terms of zero order in e do not vanish (see §4.4).

An indirect effect of the eclipses is due to the fact that, since they stop solar heating, if the satellite's thermal inertia is not too large the temperature distribution is drastically modified, and so is the force due to anisotropic thermal emission (see §5.2). This force in general is not directed along S, therefore when eclipses occur the average of its T component is not zero even for a circular orbit; this means that the interplay of eclipses and anisotropic thermal emission can produce a significant perturbation. The time needed for relaxation of temperature gradients is of the order of $\mathcal{R}^2 q/\chi$ (q, $\mathcal{R}$

and χ are the specific heat per unit volume, size and thermal conductivity of the body respectively), as can be easily shown from simple dimensional arguments. For a metallic sphere of the size of LAGEOS, this results in a relaxation time of the order of 10^3 s, i.e. of the same order as the typical duration of an eclipse. Thus, for LAGEOS, the effect is not negligible. Since for LAGEOS $\alpha \Delta \mathcal{T}/\mathcal{T}_0$ (see equation (5.28)) is of the same order as the eccentricity, at least as regards long-term semi-major axis changes, this effect might be comparable to the total effect of eclipses. To get an order-of-magnitude estimate of the resulting averaged T component, we can use equation (5.27) and divide the magnitude of the acceleration A_w by a factor of the order of 30 (15 to account for the fractional time spent in the shadow times 2 due to projection on the tangential direction). We obtain another perturbing force of the order of 10^{-11} cm s^{-2} contributing to the long-term evolution of LAGEOS' orbit, with typical frequencies depending again on linear combinations of $\lambda_\odot$ and Ω.

6
Drag

If a satellite orbits not too far from the Earth's surface it experiences a drag due to interaction with the particles of the atmosphere and/or plasmasphere. The study of the perturbative effects of drag on low-flying satellites has a long history, this being the first case of non-gravitational perturbation studied at the beginning of the space age (see Chapter 1). We do not aim at a complete presentation of this subject, not only because it would be too long, but also for two other reasons. First, there is already a very good textbook on the subject by King-Hele (1964) (this is now out of print, but a revised edition is in preparation) and there is no need to duplicate it. Second, if a satellite experiences such a strong drag that an accurate modelling of its effects becomes necessary, then a very accurate orbit determination becomes impossible; the tracking data will instead contain significant information on the drag itself, hence on the density and motions of the upper atmosphere, and this is the way they are mainly used today. On the contrary the increase in the accuracy of both satellite tracking and orbital analysis has raised interest in the modelling of the very small drag effects which can be experienced by very high-flying satellites such as LAGEOS.

This chapter contains only a few essential formulae on the orbital perturbations produced by a drag-like force (§6.1); discusses the way in which the perturbing acceleration can be calculated for a given atmospheric density (§6.2); and introduces the increased drag effect generated by charged particles (§6.3).

6.1 ORBITAL PERTURBATIONS BY A DRAG-LIKE FORCE

By 'drag-like force' we mean a force acting on the spacecraft along

ORBITAL PERTURBATIONS BY A DRAG-LIKE FORCE

the same direction of its velocity vector (in an Earth-centred, non-rotating frame such as the one introduced in §3.1); in the notation of Chapter 3, we will assume that the perturbing acceleration has only the component F_v. Moreover, the size of the force will be dependent on the velocity v of the satellite and on the density ρ of the atmosphere:

$$F_v = -\tfrac{1}{2} Z C_D \frac{\mathcal{A}}{\mathcal{M}} \rho |v| v \qquad (6.1)$$

where $\mathcal{A}/\mathcal{M}$ is the spacecraft cross sectional area (perpendicular to the velocity) divided by its mass, while the drag coefficient C_D is defined by equation (6.1) and its derivation will be discussed in §6.2. Z is a corrective coefficient accounting for the fact that the atmosphere is not at rest, but rotates more or less rigidly with the Earth and has therefore a velocity $V_A = \omega_A \times r$, with ω_A close to the Earth's angular velocity $\omega_\oplus$. The aerodynamic drag is proportional to $|v - V_A|^2 = Zv^2$ and thus Z is a function of the orbital elements too; for small eccentricities we can approximate $Z \simeq (1 - a\omega_A \cos I/v)^2$. For the purposes of this section, we will assume that

$$Z C_D \frac{\mathcal{A}}{\mathcal{M}} \equiv D \qquad (6.2)$$

is a constant.

The rate of change of the orbital elements due to a drag-like force can be computed by the formulae of Chapter 3: for the semi-major axis we obtain from equation (3.56)

$$\dot{a} = -\left(\frac{v}{na}\right)^3 na^2 \rho D; \qquad (6.3)$$

and for the eccentricity, from equation (3.57),

$$\dot{e} = -(e + \cos f) v \rho D. \qquad (6.4)$$

To compute the secular perturbations we have to average over one revolution; however, it is simpler to express the resulting integrals in terms of the eccentric anomaly u. By the energy integral (equations (3.17) and (3.19)) we get:

$$v^2 = \mu\left(\frac{2}{r} - \frac{1}{a}\right) = n^2 a^2 \frac{1 + e \cos u}{1 - e \sin u}. \qquad (6.5)$$

To change the variable in the integrals we use equation (3.25):

$$\frac{da}{du} = \frac{\dot{a}}{\dot{u}} = \frac{\dot{a}}{u}(1 - e \cos u). \tag{6.6}$$

By substituting into equation (6.3) and integrating over one orbital period $P = 2\pi/n$ we obtain for the increment of a over one orbit

$$\Delta a = -Da^2 \int_0^{2\pi} \frac{(1 + e \cos u)^{3/2}}{(1 - e \cos u)^{1/2}} \rho \, du. \tag{6.7}$$

A similar computation for e, using also equation (3.23), gives

$$\Delta e = -Da(1 - e^2) \int_0^{2\pi} \frac{(1 + e \cos u)^{1/2}}{(1 - e \cos u)^{1/2}} \rho \cos u \, du. \tag{6.8}$$

From equations such as (6.7) and (6.8) we can already draw conclusions on the qualitative evolution of the orbit under the influence of drag. First, $\Delta a < 0$, that is the satellite reduces its semi-major axis and its orbital period; this could easily have been anticipated by considering that a dissipative force must decrease the total energy (which is proportional to $-1/a$). Second, for every reasonable density distribution (provided only it is a decreasing function of the radius r), $\Delta e < 0$, unless e is 0 from the beginning. This can be seen from the integral in equation (6.8), with the integrand positive for $-\pi/2 < u < \pi/2$ and negative for $+\pi/2 < u < 3\pi/2$: however, both ρ and the factor $(1 + e \cos u)^{1/2}/(1 - e \cos u)^{1/2}$ are larger in the section of the orbit closer to perigee than in the corresponding points with $|u| > \pi/2$. This corresponds simply to the fact that if the greatest drag is experienced near perigee, the satellite does not swing out so far on the opposite side of the orbit, and as a result the maximum altitude is reduced without affecting too much the perigee distance itself.

To further discuss the effects of drag we must introduce some atmospheric model describing ρ as a function of r and possibly other parameters. The simplest such model describes ρ as an exponential function of r:

$$\rho(r) = \rho_0 \exp\left(\frac{a_0 - r}{\mathscr{H}}\right) \tag{6.9}$$

where $\mathscr{H}$ is the 'scale height', over which the density decreases by a factor $\exp(1)$ with respect to the density ρ_0 at the initial value a_0 of r. Since equation (6.9) gives the solution for the density distri-

ORBITAL PERTURBATIONS BY A DRAG-LIKE FORCE

bution of an isothermal column of gas in equilibrium with its own weight (Boltzmann's law), it represents a good approximation whenever the temperature of the atmosphere varies little with height (as it occurs in the upper atmosphere above 250 km; note that a measurement of the scale height from orbital data allows us to estimate the temperature itself). Within this model, we can give a simple formula for the decay of a satellite in a circular orbit under the effect of drag. In this case $v = na$, $r = a$ and by substitution of equation (6.9) into (6.3) we obtain

$$\dot{a} = -\rho_0 D n a^2 \exp\left(\frac{a_0 - a}{\mathcal{H}}\right). \quad (6.10)$$

To compute an explicit solution to equation (6.10) we will introduce the following approximation: $na^2 \simeq n_0 a_0^2$, a_0 nd n_0 being the values of a and n corresponding to the initial conditions. This does not introduce a very significant error, provided $\mathcal{H}$ is not too large. Then equation (6.10) can be explicitly integrated:

$$t - t_0 = \left[1 - \exp\left(\frac{a - a_0}{\mathcal{H}}\right)\right] \mathcal{H}/D\rho_0 n_0 a_0^2. \quad (6.11)$$

As a useful order-of-magnitude estimate, the time needed for a to decrease by one scale height $\mathcal{H}$ is

$$\Delta t_{\mathcal{H}} = \frac{0.31}{n_0} \frac{\mathcal{H}}{a_0} \left(\frac{n_0^2 a_0}{D\rho_0 n_0^2 a_0^2/2}\right). \quad (6.12)$$

The expression in the bracket is the ratio between the monopole term of the Earth's attraction and the drag force at $r = a_0$. The exponential atmosphere model also allows us to derive explicit relationships between the variations of different orbital elements (e.g. eccentricity and perigee), which contain the scale height as a parameter but are independent of the size and shape of the satellite or of variations in the drag. As a consequence, the scale height can be directly obtained in a quite reliable way.

Of course in the real atmosphere a more complex structure is present, and the scale height changes with r (generally speaking it increases with r). Models taking this into account can be used (see King-Hele 1964, Chap. 6), but in this case there is usually little need for a very refined analytic theory because this is only one of the problems of atmospheric models. To predict the drag experienced by a given satellite the main problem is the change of ρ with time that is mostly correlated with solar and geomagnetic activity. As a result of these external (and quite unpredictable) excitations, the atmospheric density is known to change by more than one order of magnitude at heights of a few hundreds of kilometres. Significant day–night, seasonal and latitudinal variations are there as well. As a result, predictions of drag can be considered only as order-of-magnitude estimates; an *a posteriori* analysis using solar and geomagnetic activity indices as input can be somewhat more accurate, but still not enough for very accurate orbit determination with satellites flying lower than, say, 1000 km, unless very short arcs or drag-free technology are used.

However, this does not mean that useful geodetic information cannot be extracted by orbits strongly perturbed by atmospheric drag: but we must take into account that some of the orbital elements, such as a and e, undergo a secular change which is not well determined. Periodic signals with a known frequency can be extracted in an accurate way even in presence of a secular change. For instance, resonant geopotential harmonics produce effects on the semi-major axis that can be recovered even from the tracking data of very low satellites (King-Hele and Walker 1982). Moreover, there are elements that do not undergo secular perturbations due to drag—as an example, with a drag-like force of the type given by equation (6.1) and with ρ an arbitrary function of r, there is no secular perturbation in ω. This can be proved by substituting in equation (3.47) $\dot{\Omega} = 0$ (because $W = 0$) and the expressions for T, R computed from equation (3.55):

$$\dot{\omega} = -\frac{D\rho v}{2e}\left(\sin f + \frac{e}{\beta}\cos f \sin u\right). \qquad (6.13)$$

Since $\dot{\omega}$ is an odd function of the anomalies, it averages to zero over one orbit.

Even simpler is the case of the inclination I and the nodal argument Ω: with a drag-like force acting only in the orbital plane, the out-of-plane component W is zero and the orbital plane does not change. This mathematical result unfortunately relies on physically unrealistic assumptions. Even for a spherical satellite experiencing no aerodynamic lift forces, the rotation of the atmosphere causes the relative velocity of the spacecraft with respect to it to be out of the orbital plane. Taking as an example a circular inclined orbit, the out-of-plane component of the drag is approximately

$$W = F_v \frac{\omega_A}{n} \frac{\sin I \cos \lambda}{Z^{1/2}} \qquad (6.14)$$

where Z is the correcting factor already used in (6.1) and λ is the argument of latitude (for a circular orbit, the angle between the current position and the ascending node). By using equations (3.43) and (3.40) we obtain:

$$\dot{\Omega} = \frac{F_v}{na} \frac{\omega_A}{n} \frac{1}{Z^{1/2}} \cos \lambda \sin \lambda + O(e) \qquad (6.15)$$

$$\dot{I} = \frac{F_v}{na} \frac{\omega_A}{n} \frac{1}{Z^{1/2}} \cos^2 \lambda + O(e). \qquad (6.16)$$

From the latter formulae it is easy to appreciate that the secular effect on Ω is zero for $e = 0$ (that is, it contains the factor e, and is very small for small eccentricities). On the contrary, the effect over one orbit does not vanish for I:

$$\Delta I = \pi \left(\frac{F_v}{n^2 a}\right) \frac{\omega_A}{n} \frac{\sin I}{Z^{1/2}} + O(e). \qquad (6.17)$$

As a result, the orbital plane tends to approach the equatorial plane; the term in brackets is again the ratio of the drag force to the Earth's attraction.

As an example, for a satellite of the Starlette class, F_v is about 5×10^{-8} cm s^{-2} (assuming a high level of solar and geomagnetic activity), $\mathcal{A}/\mathcal{M} \simeq 0.01$ cm^2 g^{-1} and $\Delta I \simeq 8 \times 10^{-7}$ arcsec, that is the reference system stability is not significantly degraded by drag effects. On the other hand, for closer satellites (starting from Sputnik 2; see Merson *et al* (1959)) a measurement of the rate of decrease of the inclination can yield information on the atmospheric rotation rate, which has been shown to be significantly greater than $\omega_\oplus$ due to the presence of strong west-to-east winds (King-Hele and Walker 1983).

6.2 DRAG COEFFICIENTS

The mean free path of the gas molecules in the upper atmosphere ranges from some 100 m at 180 km height to more than 1 km above

300 km. In the vicinity of a spacecraft surface, the mean free path can decrease by up to an order of magnitude if molecules are not 'elastically' reflected but re-emitted with a lower velocity (corresponding, for instance, to the surface temperature of the body). However, the mean free path remains in most cases much larger than the typical linear dimensions of satellites which as a consequence experience essentially a free molecule flow. This implies that collisions between molecules can be neglected when one investigates aerodynamic drag, and what must be analysed in detail is only the mechanism of interaction between 'single' molecules and the body surface.

An important difference between the behaviour of drag experienced by satellites at low and high altitude is connected with the ratio between the orbital velocity v and the average thermal molecular speed V_T. For low orbits $v \simeq 7.5$ km s^{-1} while V_T is close to 1 km s^{-1} for oxygen atoms at a temperature of about 10^3 K (lighter species are much less abundant at heights of a few hundred kilometres), so that the satellites move essentially in a stationary fluid with the molecules impinging only against their front surface. But at higher altitudes the situation changes, because the orbital velocity becomes smaller and smaller while V_T increases rapidly due to higher temperatures and lower mean molecular weight; and it is clear that the drag force would vanish at v/V_T approaching zero because of the isotropy of the resisting medium. We shall follow Afonso *et al* (1985) in deriving the dependence of the drag force on the ratio v/V_T for a spherical satellite, both to illustrate the essential features of the phenomenon and to get a model suitable for application to LAGEOS.

The molecule-to-spacecraft relative velocity V is connected with the molecule's thermal velocity V_T and with the orbital velocity v by the following relationships:

$$V = V_T - v \qquad V^2 = V_T^2 + v^2 - 2vV_T \cos \theta \qquad (6.18)$$
$$V \cos \Psi = V_T \cos \theta - v \qquad V \sin \Psi = V_T \sin \theta$$

where θ is the angle between v and V_T and Ψ is the angle between v and V. If the satellite absorbed all the impinging molecules, dN being the number of them with thermal speeds in the range (V_T, $V_T + dV_T$), the drag force due to these dN molecules would be

DRAG COEFFICIENTS

directed along $-v$ (for obvious symmetry reasons) and given by

$$F_v = \frac{m \, dN}{2} \frac{\mathscr{A}}{\mathscr{M}} \int_0^\pi v \cos \Psi \sin \theta \, d\theta \qquad (6.19)$$

where m is the molecular mass and we have assumed that the distribution of thermal velocities is isotropic. Computing the integral by means of equations (6.18), we obtain

$$F_v = \frac{m \, dN}{2} \frac{\mathscr{A}}{\mathscr{M}} v^2 \left(\frac{8}{3} \frac{V_T}{v} + \frac{8}{15} \frac{v}{V_T} \right) \qquad \text{for } v \ll V_T \qquad (6.20a)$$

and

$$F_v = \frac{m \, dN}{2} \frac{\mathscr{A}}{\mathscr{M}} v^2 \left[2 + \frac{4}{3} \left(\frac{V_T}{v}\right)^2 - \frac{2}{15} \left(\frac{V_T}{v}\right)^4 \right] \qquad \text{for } v \gg V_T.$$
$$(6.20b)$$

It is interesting to note that if $v \gg V_T$, we recover the usual 'hydrodynamic' force proportional to v^2; on the other hand, if $V_T \gg v$ the force is proportional to the product vV_T, thus becoming much larger than predicted by the 'hydrodynamic' formula.

So far we have assumed that the satellite is a perfect absorber; it is easy to show that for the spherical shape the same result would also have been obtained assuming a specular reflection of molecules. On the other hand, if all the molecules were re-emitted in the opposite direction from that in which they came, a drag force twice as large would arise. In order to show what happens for more realistic types of interaction between the surface and the particles, it is useful to note that for impact velocities of several kilometres per second the gaseous molecules are neither likely to be 'trapped' on a solid surface nor to sputter it, but are usually re-emitted at a lower velocity and in a partially directional way. The degree of energy exchange is usually expressed in terms of the so-called accommodation coefficient c_a, which is defined as the ratio between the energy change experienced by the impinging molecules and the maximum energy change that could take place (the latter is not equal to the total initial kinetic energy if the surface temperature is not zero). Let us call E_i, E_r and E_t the average kinetic energy of an impinging molecule, of a re-emitted molecule and of a hypothetical molecule re-emitted with a velocity corresponding to the surface temperature respectively. Then the accommodation coefficient is

$$c_a = \frac{E_i - E_r}{E_i - E_t}. \tag{6.21}$$

The ratio $\varkappa$ between the speed of a re-emitted molecule and that of an incident one is given by

$$\varkappa = \left(\frac{E_r}{E_i}\right)^{1/2} = \left[1 + c_a\left(\frac{E_t}{E_i} - 1\right)\right]^{1/2} \simeq (1 - c_a)^{1/2} \tag{6.22}$$

where the latter approximation is valid provided the kinetic temperature of incidence is much larger than the surface temperature (which is what usually happens). Now we can estimate how the drag force behaves for a given shape of the satellite and for a given re-emission law with respect to the case of total absorption. In particular we shall consider a diffuse re-emission, with the same angular dependence we defined in §4.1 for diffusion of light. It is easily shown that for a spherical body the drag is increased with respect to pure absorption by a factor

$$b = 1 + \tfrac{4}{9}\varkappa \simeq 1 + \tfrac{4}{9}(1 - c_a)^{1/2} \tag{6.23}$$

where the factor 4/9 is the same that we obtained in equation (4.82) for the radiation pressure on a diffusing sphere, while $\varkappa$ appears to account for the reduced velocity of the re-emitted molecules.

How can we estimate the accommodation coefficient which appears in equation (6.23) and in any other similar expression valid for other shapes or re-emission laws? Assume that both the incident and the surface molecules are smooth and hard spheres and that every collision occurs head-on: then from elementary mechanics the energy exchange depends only on the projectile-to-target mass ratio μ_i (i.e. on the mass ratio between an incident molecule and a surface one) and is given by

$$c_a = \frac{4\mu_i}{(1 + \mu_i)^2}. \tag{6.24}$$

Obviously, this expression represents an oversimplification for several reasons: e.g. oblique collisions; multiple interactions with surface molecules; structure of the surface lattice; dependence on the incidence angle with respect to the surface; and quantum-mechanical effects. A review of different theoretical models presented by Cook (1965) leads to the conclusion that equation (6.24) gives the correct dependence on μ_i, but the numerical value of c_a is probably overestimated by at most a factor two (and in most cases less). At any rate, since $0 < c_a < 1$ by definition, we can see from equation (6.23) that in the case of a diffusing sphere we have always $1 < b < 1.44$, i.e. the drag force can be estimated with an uncertainty of the order of 10% as a result of the complex interaction between the gas and the surface. In terms of the traditional drag coefficient C_D, already introduced in §6.1, we have

$$C_D = b\left[\frac{8}{3}\left\langle\left(\frac{V_T}{v}\right)\right\rangle + \frac{8}{15}\left\langle\left(\frac{v}{V_T}\right)\right\rangle\right] \quad \text{for } v \lesssim V_T \quad (6.25a)$$

and

$$C_D = b\left[2 + \frac{4}{3}\left\langle\left(\frac{V_T}{v}\right)^2\right\rangle - \frac{2}{15}\left\langle\left(\frac{V_T}{v}\right)^4\right\rangle\right] \quad \text{for } v \gtrsim V_T \quad (6.25b)$$

where the angular brackets mean that the corresponding quantities must be integrated over Maxwell's speed distribution.

As regards LAGEOS, Afonso et al (1985) have presented a thorough discussion of the relevant physical quantities. To date, no direct measurement of atmospheric properties is available at a height of 6000 km; the density of atomic hydrogen, which is the main constituent of the neutral thermosphere, can be estimated only by extrapolating the data from in situ measurements at lower altitudes, using for instance an exponential density distribution (see equation (6.9)). According to Afonso et al (1985), at LAGEOS' height the number density of hydrogen atoms is in the range from 3×10^3 to 1.5×10^4 cm^{-3}, while the temperature lies in the range 800 to 1200 K. These are only order-of-magnitude estimates, because large variations depending on the Sun–Earth geometry and the level of solar activity certainly occur. Thus for LAGEOS we get $\langle V_T/v \rangle \simeq 0.8$ and by equations (6.24), (6.23) and (6.25), $c_a \simeq 0.1$, $b \simeq 1.4$ and $C_D \simeq 4$ instead of the value 2.2 currently used at lower heights. The resulting drag, although quite uncertain and highly variable, is about one order of magnitude less than the observed average deceleration of LAGEOS ($\sim 3.4 \times 10^{-10}$ cm s^{-2}).

In the case of non-spherical satellites the situation is more complex. For a given orientation of the spacecraft with respect to the orbital velocity vector v, not only must the drag coefficient be computed with some assumption about the accommodation coefficient and the re-emission mechanism (some simple examples have been worked out by Schamberg (1959)), but the cross sectional area must also be estimated. Let us consider a simple but not unrealistic case, that is a cylindrical satellite rapidly spinning about its axis of maximum momentum of inertia (in this configuration the energy is a minimum for a given angular momentum of rotation, and thus an uncontrolled satellite will relax sooner or later to this rotational state owing to small external torques or to internal energy dissipation). For a length, l, to diameter, d, ratio greater than about two, the spin axis will be perpendicular to the axis of symmetry, and therefore two 'extreme' modes of motion with respect to the velocity vector are possible: spinning 'propellerwise' and tumbling end over end. In general the spin axis may be inclined at any angle with respect to v, and the 'effective' cross-sectional area is approximately given by the mean of the two 'extreme' cases:

$$\mathcal{A} = \frac{1}{2}\left[ld + \frac{2}{\pi}\left(ld + \frac{\pi d^2}{4}\right)\right] = ld\left(0.82 + 0.25\,\frac{d}{l}\right) \quad (6.26)$$

where the second addendum is the average cross section when the satellite tumbles end over end. It is interesting to note that for $l/d < 5$, $\mathcal{A}$ never differs from the maximum value (ld) by more than 13% (6% for $l/d = 2$). A similar conclusion has been shown by Cook (1965) to hold for C_D, at least with simple assumptions about the re-emission law. Moreover, the same can be said for $l/d \gtrsim 2$, when the axis of maximum momentum of inertia is the axis of symmetry. This example shows clearly that for spinning satellites the use of average values of the cross sectional area and of the drag coefficient is not likely to cause errors in the estimated force larger than the intrinsic uncertainty in C_D. Both $\mathcal{A}$ and C_D will undergo a slow oscillation as the orientation of the orbital plane and the direction of the spin axis change, but the amplitude of this oscillation will not normally exceed about 10%.

The problem is even more complicated for satellites which are three-axis stabilised and/or have large orientable wings or solar panels. Since during one orbit the spacecraft configuration (and in particular the orientation with respect to v) is likely to change, in this case 'instantaneous' values of C_D and $\mathcal{A}$ should be derived from the (supposedly known) geometrical and attitude parameters;

without detailed information on the spacecraft configuration, it is very difficult to obtain reliable estimates of the drag force. Moreover, the basic assumption that the perturbative force is directed along $-v$ can break down, due to significant transverse components ('lift' effects) which may not average out. For instance let us consider a flat plate (a wing) whose normal is directed at an angle φ from the direction of the atmospheric flux, and assume again that re-emitted molecules follow the usual diffusion law. By recalling the results of §4.1 about radiation pressure on a flat surface element (for pure diffusion) and introducing the factor $\varkappa$ when accounting for the momentum transfer due to re-emitted particles, it is easy to show that the lift force experienced by the wing (perpendicular to v) is proportional to $2\varkappa \sin \varphi \cos \varphi/3$; on the other hand the drag-like force (directed along $-v$) is proportional to $\cos \varphi (1 + 2\varkappa \cos \varphi/3)$. For $\varphi = 45°$, the ratio of the lift to the drag force is about 0.31 for $c_a \simeq 0.1$, and 0.13 for $c_a \simeq 0.9$ (a more plausible estimate when heavy species like O or N_2 are the most abundant). This confirms that for active and/or stabilised satellites the usual treatment of the drag effects can prove inadequate, and other phenomena of great complexity may show up in the orbital evolution. In these cases no quantitative modelling effort is likely to be successful, at least as far as an accurate orbit determination or prediction is needed.

6.3 CHARGED PARTICLE DRAG

As a consequence of the geomagnetic field, the Earth is surrounded by the so-called plasmasphere, a zone where charged particles are trapped to move under the influence of the magnetic force, and their number density is significantly higher than in the interplanetary space. Artificial satellites moving in this medium tend to become electrically charged, because of collisions with electrons and ions (since electrons are lighter, their speed is higher and collisions with them are more frequent even if the plasma is globally neutral) and also because of the photoelectric effect caused by solar radiation, which depletes solid surfaces of electrons. On the other hand, the collision cross section of a charged satellite (or equivalently its drag coefficient) is not the same as it would be with no charge, because the long-range electromagnetic interaction with the plasmasphere particles can affect the trajectories of the particles in a way depending on their charge-to-mass ratio and on their average speed. For instance, the motion of particles having a charge opposite to that of the satellite will be deflected in such a way to make collisions more likely; moreover, a charged satellite can exchange momentum with the particles even when there is no collision, but only a scattering process. In the end, a stationary state will be generated, in which the collisional frequencies will adjust themselves (taking into account also the photoelectric effect) in such a way as to generate a constant electric charge on the satellite. But then the orbital motion will be affected by a drag-like perturbation which cannot be modelled by the usual techniques that are applied to 'neutral' drag, as described in the previous section.

The need for models of charged particle drag is particularly acute in cases like that of LAGEOS, which moves in a region of the plasmasphere where charged particles are quite abundant (the density of protons being comparable with that of hydrogen atoms) and for which the tracking data indicate a semi-major axis decay significantly larger than any prediction based on neutral drag mechanisms. Several recent studies (Afonso *et al* 1980, 1985, Mignard 1981, Rubincam 1982, Barlier *et al* 1986) have dealt with the problem of modelling charged particle drag and presented quantitative estimates for the case of LAGEOS. In the remaining part of this section we shall give only a brief outline of these models, stressing at the same time the main uncertainties and problems which are currently being investigated.

The first step is that of obtaining an estimate of the electrostatic potential acquired by the satellite, which is proportional to its total charge. Although in the case of LAGEOS the problem is simplified by the spherical shape of the body and by the fact that the photoelectric effect is negligible, it is very difficult to predict the (negative) equilibrium potential generated by the different collision rates of electrons and ions. The electric current collected by the satellite depends on its potential, which in turn depends on the current and on the physical properties of the plasma. Thus the problem is quite complicated, even ignoring the fact that our knowledge of plasma properties is very poor. Therefore we can give only a rough estimate of the electrostatic potential U_0, deriving it for instance by Al'pert's (1974) theory or by a few available *in situ* measurements, which agree in giving average values ranging from minus one to minus a few volts. We stress however that the potential may be subjected to large variations, e.g. during eclipses, which could cause correspondingly considerable changes in the intensity of the drag-like force.

The second step is the determination of the electric potential in the region surrounding the satellite, using U_0 as a boundary condition. Around the spacecraft the plasma is not neutral, since electrons are repelled and ions are attracted by it; the positive net charge can be estimated by assuming that ions are not significantly affected (because of their large mass), while electrons are distributed according to Boltzmann's law, i.e. with a number density n_e proportional to $\exp(e_0 U/k_B T_p)$, where U is the spherically symmetric potential, e_0 is the electronic charge, k_B is Boltzmann's constant and T_p is the plasma temperature. The solution of Poisson's equation yields U as a function of r, the distance from the satellite's centre. This solution depends critically on the parameter L_D, the so-called Debye length given by $(k_B T_p/n_e e_0^2)^{1/2}$. L_D is the scale over which the potential approaches a constant value, as a result of the fact that the negative charge on the satellite becomes balanced by the excess of positive charge in its vicinity. When $L_D \to \infty$, the solution for $U(r)$ becomes simply a Coulomb potential, because in this case the plasma is not significantly affected by the presence of the satellite.

Once the potential $U(r)$ is known, the collision cross section of the satellite can be computed by studying the trajectories of particles starting from infinity with some impact parameter b_i and some energy E_∞, and determining whether their distance of closest approach to the centre of the satellite is smaller or larger than its radius. Although the trajectories are complex and numerical integrations are needed to determine their exact behaviour (see Afonso *et al* 1985), an order-of-magnitude estimate of the satellite's cross section can be obtained by using the energy integral to evaluate the maximum value of b_i resulting in zero radial velocity at a distance less than $\mathcal{R}$. From this simplified approach one finds that the 'geometrical' cross section $\pi \mathcal{R}^2$ is increased by a factor $(1 - e_0 U_0/E_\infty)$. In the case of LAGEOS this factor is about three, and since the plasma temperature is of the order of 5000 K (with perhaps a factor of variability of two), the drag coefficient C_D is increased from the 'neutral' value of about 7, that follows from equations (6.25), to about 20. Thus the drag effect due to charged particles is amplified with respect to the case of the neutral atmosphere both because of the higher thermal velocity and because of the increased collision cross section due to interaction between the satellite and the plasma. The estimate given above can be considered fairly reliable, since it agrees satisfactorily with the experimental results by Knetchel and Pitts (1965) and with the detailed theoretical study by Jastrow and Pearse (1957). All these investigations have also shown that the contribution to the drag force of the particles which do not collide, but are only deviated by the satellite's field (thus exchanging linear momentum with it), is not very important, being in the case of LAGEOS no more than 10% of the total force.

At the altitude of LAGEOS the current knowledge of the plasma properties (in particular of the density and temperature) is very limited, and moreover these quantities depend strongly both on geometrical parameters like latitude, solar hour and declination of the Sun, and on the level of solar and geomagnetic activity, thus making it quite likely that there will be complex variations of the drag intensity on different time-scales. The present state of the art is not such that detailed models of these effects are available, even if there is no doubt that charged particle drag provides a substantial part of the observed drag-like force on LAGEOS. For instance, the implied density of H^+ ions is of the order of 10^4 cm^{-3}, a number consistent with the (few) available data. Further investigations on this problem will certainly be carried out during the years to come with the purpose of modelling the force in such a way to prevent the corresponding rapid degradation of orbital predictions and analyses performed in the frame of space geodesy experiments. But it is possible that in the future the drag force inferred by tracking data will be used as a precious source of information about the physical state of the Earth's plasmasphere.

7

MANOEUVRES

7.1 ORBITAL MANOEUVRES

Earth-based control centres usually perform orbital manoeuvres of artificial satellites for various reasons. For instance, satellites launched for monitoring Earth resources are manoeuvred in order to maintain a given optimised dynamical configuration, with respect to both the Earth and the Sun, which would otherwise be destroyed by perturbative effects. Geosynchronous satellites (both telecommunication and meteorological satellites) undergo orbital manoeuvres in order to prevent their running away from the longitude window of interest for the country which launched them. From the point of view of scientists interested in recovering relevant geophysical information by comparison of the predicted orbit with the tracking data, orbital manoeuvres simply mean an interruption of the experiment. In particular, if the main purpose of an experiment is to measure some parameter whose effect on the satellite's orbit is small but accumulates with time, the accuracy attainable with that experiment can be seriously degraded owing to orbital manoeuvres. This also means that scientists must be able to estimate how often the manoeuvres will be performed in the particular case of the experiment that they are trying to carry out.

As an example, we are going to show how this can be done, using the tools developed so far in this book, in the particular case of a geosynchronous satellite (having zero eccentricity and zero inclination on the Earth's equatorial plane). The satellite mean longitude λ_G with respect to Greenwich (i.e. in a reference system rotating with the Earth's angular velocity $\omega_\oplus$) can be written—in analogy with (4.31)—as the sum of two terms

$$\lambda_G = \lambda - \omega_\oplus t = \lambda_{Gi} + \varepsilon \tag{7.1}$$

where λ_{Gi} is the mean longitude (with respect to Greenwich) of an ideal geosynchronous satellite moving with the osculating mean motion at any time. We shall concern ourselves only with the λ_{Gi} term, because we know that its second derivative is proportional to the perturbing force and therefore it will grow quadratically with time whereas the effect in ε accumulates linearly. The acceleration $\ddot{\lambda}_{Gi}$ is related to $\dot{a}$ through the equation

$$\ddot{\lambda}_{Gi}(t) = -\frac{3}{2}\frac{n_0}{a_0}\dot{a}(t) \tag{7.2}$$

which like equation (4.36) is valid to the first perturbative order. Here $n_0 = \omega_\oplus$ and a_0 are the (constant) mean motion and semi-major axis of the geosynchronous satellite moving in absence of perturbations. But when a perturbation is present, the semi-major axis can change according to the Gauss equation

$$\dot{a}(t) = \frac{2}{n_0} T + O(e) \tag{7.3}$$

which is valid to zero order in the eccentricity and to the first perturbative order. The along-track perturbing acceleration T in the case of a geosynchronous satellite is mostly due to the asphericity of the Earth, and can be obtained from the perturbing potential in the form (2.1) by assuming zero latitude, $\lambda = \lambda_{Gi}$ and $r = a_0$:

$$T = \frac{1}{a_0}\frac{\partial V}{\partial \lambda_{Gi}} = -\sum_{l=2}^{\infty}\sum_{m=0}^{l}\frac{GM_\oplus}{a_0^2}\left(\frac{R_\oplus}{a_0}\right)^l m J_{lm} \sin[m(\lambda_{Gi} - \lambda_{lm})] P_{lm}(0) \tag{7.4}$$

which tells us that only the geopotential coefficients with $m \neq 0$ and $(l-m)$ even (since $P_{lm}(0) = 0$ for $(l-m)$ odd) perturb the orbit of a geosynchronous satellite. The largest term is the one with $l=2$, $m=2$, so that T can be approximated by

$$T = -\frac{2GM_\oplus}{a_0^2}\left(\frac{R_\oplus}{a_0}\right)^2 J_{22} \sin[2(\lambda_{Gi} - \lambda_{22})] P_{22}(0). \tag{7.5}$$

The values of the coefficients J_{22} and λ_{22} given by the most recent geopotential models are $J_{22} = 2.8075 \times 10^{-6}$, $\lambda_{22} = -14°.95$. Using

ORBITAL MANOEUVRES

the constant

$$\varkappa \equiv \frac{2GM_\oplus}{a_0^2}\left(\frac{R_\oplus}{a_0}\right)^2 J_{22} P_{22}(0) \simeq 5.6 \times 10^{-6} \text{ cm s}^{-2} \quad (7.6)$$

and the new variable $\varphi = 2(\lambda_{Gi} - \lambda_{22}) + \pi$, we have $T \simeq \varkappa \sin \varphi$, and by equations (7.2) and (7.3) we obtain the equation of motion in φ:

$$\ddot{\varphi} = -\frac{6}{a_0} \varkappa \sin \varphi. \quad (7.7)$$

This is the well-known pendulum equation. Because of the factor 2 in the definition of φ, we have four equilibrium longitudes, two stable and two unstable. The period T_{lib} of small librations about the two stable equilibrium positions is given by

$$T_{\text{lib}} = 2\pi \left(\frac{a_0}{6\varkappa}\right)^{1/2} \simeq 800 \text{ days}. \quad (7.8)$$

Therefore a satellite close to these two longitudes would not need any manoeuvre to be kept there; but of course there is the need for geosynchronous satellites at different longitudes, and in these cases manoeuvres are needed to prevent the satellite from wandering away. The dynamics governed by equation (7.7) can be visualised in the phase plane $(\varphi, \dot{\varphi})$; notice that if $\Delta a = a - a_0$,

$$\ddot{\varphi} = -3 \frac{n_0}{a_0} \dot{a} = -3 \frac{n_0}{a_0} \Delta \dot{a} \quad (7.9)$$

from equation (7.2), and therefore

$$\dot{\varphi} = -3 \frac{n_0}{a_0} \Delta a \quad (7.10)$$

since for $t = 0$, $\dot{\varphi} = 0$ and $\Delta a = 0$. Hence the phase plane gives directly the behaviour of the satellite in the plane $(\varphi, \Delta a)$, i.e. in a strip surrounding the equator at geosynchronous distance. The motion can be represented as in figure 7.1, where the curve C corresponds to an initial semi-major axis significantly larger than the geosynchronous one, resulting in a 'circulation' all around the Earth; the curve L corresponds to a librational motion around an equilibrium point; and the curve S is the separatrix between the two different kinds of motion. The orbits of geosynchronous satellites during their operational life are of the libration type. However, since the libration amplitude would in general be much larger than is required by the mission purposes, orbital manoeuvres are needed to maintain the satellite in the neighbourhood of the nominal longitude $\bar{\lambda}_{Gi}$ with a tolerance of $\pm w$ (usually $1°$ or less). The arc between two successive manoeuvres is, for a small w, an arc of parabola and its duration depends on the value of the longitude acceleration $\ddot{\lambda}_{Gi}(\bar{\lambda}_{Gi}) = \ddot{\varphi}(\bar{\lambda}_{Gi})/2$ (obtained from equation (7.7)) at the nominal longitude. The closer the satellite is to an equilibrium point, the smaller is $\ddot{\lambda}_{Gi}$ and the longer is the arc duration Δt. Since w is usually small, $\ddot{\lambda}_{Gi}$ can be assumed to be constant during the whole arc, and therefore

$$\Delta t \simeq 4 \left(\frac{w}{\ddot{\lambda}_{Gi}(\bar{\lambda}_{Gi})} \right)^{1/2} \qquad (7.11)$$

where the satellite is assumed to start at $\bar{\lambda}_{Gi} - w$, to reach $\bar{\lambda}_{Gi} + w$ and then to come back to the initial longitude. For typical values like $w = 1°$, $\ddot{\lambda}_{Gi} = 10^{-3°}$ day^{-2}, Δt is of the order of a few months. If the tracking data are used to determine the parameters J_{22} and λ_{22}, the attainable accuracy will depend not only on the accuracy of the available data, but also on the duration of the orbital arc. All the perturbations with a period much shorter than Δt can be quite easily averaged out in the data analysis, so that their intrinsic uncertainty contributes to the final uncertainty in a way inversely proportional to Δt (this is for instance the case with the short-

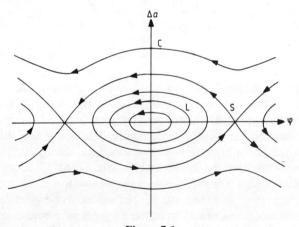

Figure 7.1

periodic solar radiation perturbations of period of the order of one day). However, longer orbital arcs will not help when dealing with long-periodic or secular perturbations (e.g. due to solar radiation pressure); their effect will accumulate in the same way as the geopotential effect and the accuracy of the measurement will depend mainly on the accuracy of the perturbation model.

7.2 ATTITUDE MANOEUVRES AND THRUSTER LEAKAGES

Most active satellites are subject not only to manoeuvres changing their orbital motion, but also to manoeuvres changing their attitude or their spin rate owing to operational constraints. The latter manoeuvres usually involve smaller thrusts than the orbital ones, but are performed more frequently. They would cause no orbital change only provided pairs of thrusters were used with a perfect thrust balance in such a way as to produce a zero net force on the spacecraft. The intensity of thrusts can be approximately balanced by alternately operating the two thrusters of a pair; but if the balance is not perfect, or if some misalignment of the thrusters is present, this results into a small net force affecting the orbit.

To obtain some orders of magnitude, let us consider a spin-stabilised cylindrical satellite of mass $\mathcal{M}$, radius $\mathcal{R}$, rotation rate ω_s, controlled by a pair of lateral thrusters with a typical thrust intensity F, and whose axes are misaligned by a small angle γ. Then the time required by the manoeuvre to change the attitude by $\Delta\theta$ is

$$\Delta t_M = \frac{1}{4} \frac{\mathcal{M}\mathcal{R}\omega_s \, \Delta\theta}{F} \qquad (7.12)$$

since the torque is $2F\mathcal{R}$ and the required angular momentum change is $\mathcal{M}\mathcal{R}^2 \omega_s \, \Delta\theta/2$. The resulting change in the orbital velocity of the satellite due to thruster misalignment will be

$$\Delta v_M = \frac{2F \Delta t_M \sin \gamma}{\mathcal{M}} = \tfrac{1}{2}\omega_s \mathcal{R} \, \Delta\theta \, \sin \gamma. \qquad (7.13)$$

For a manoeuvre aiming at changing the spin rate by $\Delta\omega_s$, equations (7.12) and (7.13) can be applied once $\Delta\theta$ is substituted by

$\Delta\omega_s/\omega_s$. It is worth noting that in both cases Δv_M does not depend on $\mathcal{M}$ and F. For typical values like $\mathcal{R} = 10^2$ cm, $\omega_s = 1$ rad s^{-1} and $\gamma = 10^{-2}$ rad, a manoeuvre changing the attitude by 0.1 rad or the spin rate by 0.1 rad s^{-1} will impart a velocity change Δv_M of 0.05 cm s^{-1}, which for an Earth satellite orbiting at several kilometres per second implies changes in the orbital elements $\Delta a/a \sim \Delta e \sim \Delta I \sim 10^{-7}$. Of course the above statement is true when the direction of the force is random, uncorrelated with the orbital motion; in each particular case, we can get the variations of the elements by using Gauss's perturbation equations (see §3.2) where the R, T, W acceleration components are substituted by the corresponding Δv_M components (this method is rigorously valid only for impulsive forces, but is a good approximation because Δt_M is usually much shorter than one orbital period). These orbital changes are often negligible for the satellite operational purposes, but obviously this is not the case when the orbit is tracked and analysed for space geodesy experiments. It is generally impossible to predict or to model in a reliable way the orbital changes due to attitude manoeuvres, and as a consequence these manoeuvres are as effective in interrupting causally connected orbital arcs as those specifically aimed at modifying the orbit. It is very important to take this factor into account when planning or performing orbital analysis of active satellites. The situation is even worse for three-axis stabilised satellites, which must manoeuvre more frequently and have more complex attitude control systems and operation procedures. Moreover, the present trend is to automate these manoeuvres, i.e. to put them under control of the spacecraft computer, which is programmed to routinely correct the attitude and/or the spin rate whenever some threshold discrepancy is reached with respect to a predetermined optimal state. In this case, even knowing when the manoeuvres are carried out may not be straightforward.

Even in the absence of manoeuvres, leakages of the thrusters' flow-control valves may impart to the spacecraft some velocity change. Usually each valve is allowed a maximum leak rate during acceptance tests and check-out; we can assume that this threshold leak rate $\dot{\mathcal{M}}$ is of the order of 100 g yr^{-1}, that is about 3×10^{-6} g s^{-1}. Because of the low leakage rates involved and the space environment in which the thrusters operate, the propellant gas flow can be assumed to be in the free-molecular gas dynamic regime in the thrusters. Then if the nozzle expansion angles are small, the disturbance acceleration due to leakage is

$$F_l \simeq \frac{\dot{\mathcal{M}}}{\mathcal{M}} V_{th} \tag{7.14}$$

where V_{th} is the average thermal velocity of the propellant molecules. According to the principle of equipartition of energy, the average thermal velocity of a molecule of mass m_g is

$$V_{th} \simeq \left(\frac{3k_B T_g}{m_g}\right)^{1/2} \tag{7.15}$$

where T_g is the gas temperature and k_B the Boltzmann constant. The factor 3 comes from the fact that only the three translational degrees of freedom of the molecule (not the rotational ones) are relevant to compute the effect of gas leakage on the spacecraft motion. For hydrazine (N_2H_4) gas at a temperature of 200 °C, $V_{th} \simeq 0.8$ km s^{-1}; if $\mathcal{M} = 10^6$ g and $\dot{\mathcal{M}} = 3 \times 10^{-6}$ g s^{-1}, the resulting spacecraft acceleration is about 2×10^{-7} cm s^{-2}. This perturbation is not easy to model, because ground tests have shown that in general leakages change as a function of valve actuations (because of contamination of the valve seats, different seating characteristics, etc). However, if within an orbital arc the spacecraft attitude is constant, it is likely that including in the force model a constant unknown force will significantly reduce the unmodelled acceleration; moreover, from the geometry of the thruster configuration it is sometimes possible to predict the direction of the leakage force, and in this case the only remaining unknown is its magnitude. These considerations show clearly that both manoeuvres and leakage forces cannot be neglected when tracking data of active satellites are analysed. However, in many cases these factors do not completely prevent the extraction of useful information from the data, but just require a careful assessment of their relevance for a particular mission.

REFERENCES

Afonso G, Barlier F, Berger C and Mignard F 1980 Effet du freinage atmosphérique et de la traînée électrique sur la trajectoire du satellite LAGEOS *C. R. Acad. Sci., Paris* **290** 445–8

Afonso G, Barlier F, Berger C, Mignard F and Walch J J 1985 Reassessment of the charge and neutral drag of LAGEOS and its geophysical implications *J. Geophys. Res.* **90** 9381–98

Aksnes K 1976 Short-period and long-period perturbations of a spherical satellite due to direct solar radiation *Celestial Mech.* **13** 89–104

Al'pert Ya L 1974 Waves and satellites in the near-Earth plasma *Studies in Soviet Science* (New York: Consultants Bureau) pp. 45–57

Anselmo L, Bertotti B, Farinella P, Milani A and Nobili A M 1983a Orbital perturbations due to radiation pressure for a spacecraft of complex shape *Celestial Mech.* **29** 27–43

Anselmo L, Farinella P, Milani A and Nobili A M 1981 Modelling of orbital perturbations due to radiation pressure for high Earth satellites in *ESA Spacecraft Flight Dynamics* ESA SP-160 pp. 47–52

—— 1983b Effects of the Earth-reflected sunlight on the orbit of the LAGEOS satellite *Astron. Astrophys.* **117** 3–8

Arnold V I 1983 *Geometrical Methods in the Theory of Ordinary Differential Equations* (Berlin: Springer)

Ash M E, Shapiro I I and Smith W B 1971 The system of planetary masses *Science* **174** 551–6

Ashby N and Bertotti B 1984 Relativistic perturbations of an Earth satellite *Phys. Rev. Lett.* **52** 485–8

Barlier F, Carpino M, Farinella P, Mignard F, Milani A and Nobili A M 1986 Non-gravitational perturbations on the semimajor axis of LAGEOS *Ann. Geophys.* **4** A3 193–210

Burns J A 1976 Elementary derivation of the perturbation equations of celestial mechanics *Am. J. Phys.* **44** 944–9

Carpino M, Farinella P, Milani A and Nobili A M 1986 Sensitivity of

LAGEOS to changes in Earth's (2.2) gravity coefficient *Celestial Mech.* in press

Catalano S, McCrosky R, Milani A and Nobili A M 1983 Optical tracking of synchronous Earth's satellites for geophysical purposes *J. Geophys. Res.* **88** 669–76

Cook G E 1965 Satellite drag coefficients *Planet. Space Sci.* **13** 929–46

Darwin G H 1908 *Scientific Papers* Vol. 2 (Cambridge: Cambridge University Press)

Farinella P, Milani A, Nobili A M and Sacerdote F 1981 Dynamics of an artificial satellite in an Earth-fixed reference frame: effects of polar motions in *Reference Coordinate Systems for Earth Dynamics* ed E M Gaposchkin and B Kolaczek (Dordrecht: Reidel) pp. 271–4

Gube M 1982 Planetary albedo estimates from Meteosat data *ESA Journal* **6** 53–69

Hori G 1966 in *Lectures in Applied Mathematics* Vol. 7 part 3 (Providence, RI: American Mathematical Society) pp. 167–78

Jastrow R and Pearse C A 1957 Atmospheric drag on the satellite *J. Geophys. Res.* **62** 413–23

Kaula W M 1966 *Theory of Satellite Geodesy* (Waltham, MA: Blaisdell)

King-Hele D G 1964 *Theory of Satellite Orbits in an Atmosphere* (London: Butterworths)

—— 1983 Geophysical researches with the orbits of the first satellites *Geophys. J. R. Astron. Soc.* **74** 7–23

King-Hele D G and Walker D M C 1982 Evaluation of 15th-order harmonics in the geopotential from analysis of resonant orbits *Proc. R. Soc.* A **379** 247–88

—— 1983 Upper-atmosphere zonal winds, from analysis of satellite orbits *Planet. Space Sci.* **31** 509–35

Kittel C and Kroemer H 1980 *Thermal Physics* (San Francisco: Freeman)

Knechtel E D and Pitts W C 1965 Experimental investigation of electric drag on spherical satellite models *NASA Technical Note* NASA TN D-2619

Kovalevsky J 1963 *Introduction à la Mécanique Céleste* (Paris: Librairie Armand Colin)

Kozai Y 1961 *Smithsonian Astrophys. Obs. Spec. Rep.* No 56 (Washington, DC: Smithsonian Institute)

Lambeck K 1980 *The Earth's Variable Rotation* (Cambridge: Cambridge University Press)

Lerch F J, Klosko S M, Laubscher R E and Wagner C A 1979 Gravity model improvement using Geos 3 (GEM 9 and 10) *J. Geophys. Res.* **84** 3897–916

Lerch F J, Klosko S M and Patel G B 1983 A refined gravity model from LAGEOS (GEM-L2) *NASA Technical Memorandum* 84986

Levin E 1962 Reflected radiation received by an Earth satellite *ARS Journal* (Sept) 1328–31

Manakov Yu M 1977 Perturbing effect of terrestrial thermal radiation pressure on an artificial Earth satellite *Geodesy, Mapping and Photogrammetry* **18** 95–8

Merson R H, King-Hele D G and Plimmer R N A 1959 Changes in the inclination of satellite orbits to the equator *Nature* **183** 239–40

Mignard F 1981 Action de l'atmosphere neutre et ionisee sur le mouvement d'un satellite — Application a LAGEOS *Ann. Geophys.* t.37 fasc.1 247–52

Munk W H and MacDonald G J F 1960 *The Rotation of the Earth. A Geophysical Discussion* (Cambridge: Cambridge University Press)

Peters P C 1964 Gravitational radiation and the motion of two point masses *Phys. Rev.* **136** B1224–32

Raschke E and Bandeen W R 1970 The radiation balance of the planet Earth from radiation measurements of the satellite Nimbus 2 *J. Appl. Meteorol.* **9**

Roy A E 1978 *Orbital Motion* 1st edn (Bristol: Adam Hilger)

Rubincam D P 1982 On the secular decrease in the semimajor axis of LAGEOS's orbit *Celestial Mech.* **26** 361–81

Schamberg R 1959 *Rand Research Memorandum* RM-2313

Sehnal L 1981 Effects of the terrestrial infrared radiation pressure on the motion of an artificial satellite *Celestial Mech.* **25** 169–79

Smart W M 1947 J C Adams and the discovery of Neptune *Occasional Notes of the RAS* No 11 (London: Royal Astronomical Society)

Smith D E 1970 Earth-reflected radiation presure in *Dynamics of Satellites (1969)* ed B Morando (Berlin: Springer) pp. 284–94

Sterne T E 1960 *An Introduction to Celestial Mechanics* (New York: Interscience)

Wahr J 1981 The forced nutations of an elliptical, rotating, elastic and oceanless Earth *Geophys. J. R. Astron. Soc.* **64** 705

Whipple F L and Sekanina Z 1979 Comet Encke: precession of the spin axis, non-gravitational motion and sublimation *Astron. J.* **84** 1894–909

INDEX

Absorption (molecular), 104, 105, 106
Absorption coefficient, 19, 48, 74, 88
Accommodation coefficient, 105, 108, 109
Angular momentum, 28, 34, 53, 63
Antenna, 8, 55, 62, 64, 72
Apparent forces, 23–6
Area-to-mass ratio, 2, 12, 58, 93, 95, 99
Argument
  of latitude, 38, 59, 86, 103
  of perigee, 30, 35, 36, 102
Ascending node, 29, 35, 63, 103
Atmospheric
  density, 17, 99, 107
  drag, 2, 17, 98–112
  rotation, 99, 102, 103
Averaging, 44, 47, 70

Black body, 49, 88
Boltzmann constant, 111, 119

Celestial mechanics, 1, 27–47
Charged particle drag, 18, 75, 78, 109–12
Conductivity coefficient, 89

Debye length, 111
Diffusion coefficient, 48, 74
Double integration, 44
Drag coefficient, 18, 99, 103–9, 111

'Drag-free' technology, 2, 8, 102
Dynamic effect (tides), 20
D'Alembert series, 39, 57, 61

ERS-1, 8, 9, 73
Earth's
  albedo, 19, 78, 80, 84, 92
  infrared emission, 86, 92
  mass, 12, 13, 28
  oblateness, 17, 45, 55, 87
  rotation, 23, 24, 47, 72, 99, 103, 113
  shadow, 94
Earthshine, 4, 18, 47, 77–87
Eccentric anomaly, 31, 94, 99
Eccentricity, 28, 34, 56, 65, 100
Eclipse, 18, 94–7
Electrostatic potential, 110
Elliptic orbits, 28
Emissivity coefficient, 88
Energy (orbital), 30, 33, 99, 100
Equatorial reference frame, 29, 80

Fourier series, 39, 54, 61, 63, 66, 70, 82, 86

GEM geopotential models, 13, 17, 28, 114
Geostationary satellite, 10, 24, 58, 67, 113
Global Positioning System, 7, 72, 94
Gravitational
  constant, G, 12
  waves, 20

INDEX

Heat conduction, 89

Inclination, 30, 35, 40, 63, 72, 103

Keplerian
  elements, 27–32
  orbits, 27–32, 42, 43
Kepler's
  equation, 31, 32
  third law, 13, 30, 59
Kinematic effect (tides), 20

LAGEOS, 2, 6, 11, 12, 18, 24, 55, 73, 77, 80, 88, 92, 96, 107, 110, 112
Lambert's diffusion, 49, 74, 90
Laser ranging, 6
Leaks, 4, 118
Legendre functions, 13, 17, 89, 90, 114
Length of the day, 23, 24, 25
Lenz vector, 28
Lift (aerodynamic), 109
Longitude of the perigee, 40
Love number, 20, 22

Manoeuvre, 4, 68, 113–19
Mean anomaly, 31, 36, 37, 39, 46, 54
Mean free path, 103
Mean longitude, 40, 113
Mean motion, 30, 37, 59, 114
Meteosat, 84
Micrometeorite, 19
Mirror reflection, 49, 84, 105
Misalignment angle, 56, 58, 69, 93
Monopole, 12, 13, 42
Moon, 17, 20

Non-singular elements, 38–42, 65
Nutation, 23

Oceanic loading, 21, 22
Oceanic tides, 22
Osculating elements, 32
Out-of-plane perturbation, 33, 40, 53, 61, 62, 63, 71, 102

Perigee, 28, 32, 53, 100
Perturbation, 32, 42
Polar wobble, 23, 24, 72, 75
Power system, 4, 19, 94
Poynting–Robertson drag, 19
Precession (of satellite), 52

Radar altimeter, 8
Radial perturbation, 33, 53, 61, 62, 71, 79, 95
Radiation pressure, 3, 18, 48–97
Radiowave beam, 4, 7, 19, 55, 72, 93–7
Rapid variable, 44
Reference system, 20, 29
Reflection coefficient, 48, 74
Relativity, 17, 20
Residuals (in distance), 6, 7
Resonance, 46, 67, 102

SEASAT, 8, 18, 73
Scale height, 100, 101
Schwarzschild radius, 17
Seasonal effects, 79, 85, 88, 102
Secular perturbations, 42–7, 54
Semi-major axis, 28, 31, 33, 43, 53, 70, 100
Solar
  activity, 102, 107, 112
  constant, 18, 49, 52, 88
Spherical harmonics, 13, 17, 46, 67, 89, 102, 116
Spinning satellite, 52, 69, 88, 108
Sputnik, 2, 103
Starlette, 10, 18, 55, 103
Stephan–Boltzmann constant, 88
Sun, gravitation of, 17, 20

INDEX

Temperature
  of atmosphere, 101, 104, 107
  of ejected gas, 119
  of plasma, 111, 112
  of spacecraft, 19, 53, 87–93, 96, 104, 106
Terminator, 79
Thermal radiation, 4, 19, 86, 87–93, 96, 106
Tidal friction, 1
Tides, 20, 22, 23, 82

Torque, 34, 109, 117
Tracking station, 6, 20, 72, 75
Transverse perturbation, 33, 53, 61, 71, 79, 95
True anomaly, 28, 30, 54

Velocity
  of light, 17, 48, 49, 90
  molecular, 104
  orbital, 30, 37, 99, 104, 108
Venus, 17